Student Solutions Manual

for

Kaseberg's

Introductory Algebra
A Just-in-Time Approach

Third Edition

Kevin Bodden
Lewis & Clark Community College

THOMSON

BROOKS/COLE

Australia • Canada • Mexico • Singapore • Spain • United Kingdom • United States

For more information about our products,
contact us at:
Thomson Learning Academic Resource Center
1-800-423-0563

For permission to use material from this text,
contact us by:
Phone: 1-800-730-2214
Fax: 1-800-730-2215
Web: http://www.thomsonrights.com

Thomson Brooks/Cole
10 Davis Drive
Belmont, CA 94002-3098
USA

Asia
Thomson Learning
5 Shenton Way #01-01
UIC Building
Singapore 068808

Australia/New Zealand
Thomson Learning
102 Dodds Street
Southbank, Victoria 3006
Australia

Canada
Nelson
1120 Birchmount Road
Toronto, Ontario M1K 5G4
Canada

Europe/Middle East/South Africa
Thomson Learning
High Holborn House
50/51 Bedford Row
London WC1R 4LR
United Kingdom

Latin America
Thomson Learning
Seneca, 53
Colonia Polanco
11560 Mexico D.F.
Mexico

Spain/Portugal
Paraninfo
Calle/Magallanes, 25
28015 Madrid, Spain

Acknowledgements

A special thank-you to Cindy Rubash, author of the Second Edition of the Student Solutions Manual, for providing the foundation for this edition.

CONTENTS

Chapter 1

Exercises 1.1

1. For example: Assume you work 20 days a month or you work every day of the month.

3. For example: Assume your teacher knows your last name and in which class you are enrolled.

5. 3 panels for 1st pen, 2 additional panels for each pen after that.
$3 + 2 \cdot 19 = 41$

7. Dean is assuming that the ponies may be tied to the panels.
3 panels for the 1st pen, 2 additional panels for each pen after that.
$3 + 2 \cdot 19 = 41$

9. 3 panels are required for each pen.
$3 \cdot 20 = 60$

11. Answers may vary.

13. Answers may vary.

15. Draw a picture; try a simpler problem; look for a pattern; use a table of values; compare with a previous problem; make an estimate.

Exercises 1.2

1. a. $0.30 = \dfrac{3}{10}$

$0.30 = \dfrac{30}{100} = 30\%$

b. $0.05 = \dfrac{5}{100} = \dfrac{1}{20}$

$0.05 = \dfrac{5}{100} = 5\%$

c. $0.125 = \dfrac{125}{1000} = \dfrac{1}{8}$

$0.125 = \dfrac{0.125}{1} = \dfrac{0.125 \times 100}{1 \times 100} = \dfrac{12.5}{100}$
$= 12.5\%$

d. $0.02 = \dfrac{2}{100} = \dfrac{1}{50}$

$0.02 = \dfrac{2}{100} = 2\%$

3. a. $15\% = \dfrac{15}{100} = 0.15$

$15\% = \dfrac{15}{100} = \dfrac{3}{20}$

b. $0.5\% = \dfrac{0.5}{100} = \dfrac{0.5 \times 10}{100 \times 10} = \dfrac{5}{1000}$
$= 0.005$

$0.5\% = \dfrac{0.5}{100} = \dfrac{0.5 \times 2}{100 \times 2} = \dfrac{1}{200}$

c. $48\% = \dfrac{48}{100} = 0.48$

$48\% = \dfrac{48}{100} = \dfrac{12}{25}$

d. $250\% = \dfrac{250}{100} = 2.5$

$250\% = \dfrac{250}{100} = \dfrac{5}{2}$ or $2\dfrac{1}{2}$

5. $\dfrac{3}{4} \cdot \dfrac{1}{2} = \dfrac{3 \cdot 1}{4 \cdot 2} = \dfrac{3}{8} = 0.375$

7. $\dfrac{1}{3} + \dfrac{1}{2} = \dfrac{2}{6} + \dfrac{3}{6} = \dfrac{5}{6} \approx 0.833$

9. $5 \div \dfrac{2}{15} = \dfrac{5}{1} \cdot \dfrac{15}{2} = \dfrac{75}{2} = 37\dfrac{1}{2} = 37.5$

11. $\dfrac{3}{5} \cdot \dfrac{4}{9} = \dfrac{3 \cdot 4}{5 \cdot 9} = \dfrac{12}{45} = \dfrac{4}{15} \approx 0.267$

13. $2\dfrac{2}{3} - 1\dfrac{1}{4} = \dfrac{8}{3} - \dfrac{5}{4} = \dfrac{32}{12} - \dfrac{15}{12} = \dfrac{17}{12}$
$= 1\dfrac{5}{12} \approx 1.417$

15. $2\dfrac{2}{3} \div 1\dfrac{1}{4} = \dfrac{8}{3} \div \dfrac{5}{4} = \dfrac{8}{3} \cdot \dfrac{4}{5} = \dfrac{32}{15}$
$= 2\dfrac{2}{15} \approx 2.133$

17. $\dfrac{4}{5} - \dfrac{1}{3} = \dfrac{12}{15} - \dfrac{5}{15} = \dfrac{7}{15} \approx 0.467$

19. $1\dfrac{1}{2} + 2\dfrac{1}{4} = \dfrac{3}{2} + \dfrac{9}{4} = \dfrac{6}{4} + \dfrac{9}{4} = \dfrac{15}{4} = 3\dfrac{3}{4} = 3.75$

21. $1\frac{1}{2}+2\frac{1}{2}=\frac{3}{2}+\frac{5}{2}=\frac{8}{2}=4$

23. $1\frac{1}{4}-\frac{3}{8}=\frac{5}{4}-\frac{3}{8}=\frac{10}{8}-\frac{3}{8}=\frac{7}{8}=0.875$

25. Answers may vary.

27. Answers may vary.

29. Answers may vary.

31.

$-3\ -2\frac{1}{2}\qquad -\frac{1}{2}\qquad \frac{3}{4}\ \ 1.5\ \ 2\frac{1}{4}$

$-4\quad -3\quad -2\quad -1\quad 0\quad 1\quad 2\quad 3\quad 4$

33.

$\frac{8}{7}\,1\frac{1}{6}\qquad 1.25$

$1.0\quad 1.1\quad 1.2\quad 1.3\quad 1.4\quad 1.5$

35. a. Natural numbers are integers that are not negative or zero.

b. Negative real numbers could be used to write elevations below see level.

c. Rational numbers include ratios of integers and terminating or repeating decimals.

d. The even numbers and odd numbers combine to form the set of integers.

37. a. $\text{small}=1\frac{1}{2}$ yd, $\text{large}=1\frac{3}{4}$ yd

$1\frac{1}{2}+1\frac{3}{4}=\frac{3}{2}+\frac{7}{4}=\frac{6}{4}+\frac{7}{4}=3\frac{1}{4}$ yd

b. $5\cdot\left(1\frac{3}{4}\right)=5\cdot\frac{7}{4}=\frac{35}{4}=8\frac{3}{4}$ yd

c. $17\text{ yd}\div2\frac{1}{8}\text{ yd}=17\div2\frac{1}{8}=17\div\frac{17}{8}$

$=17\cdot\frac{8}{17}=\frac{136}{17}$

$=8$ shirts

d. $1\frac{3}{4}\cdot2=\frac{7}{4}\cdot2=\frac{14}{4}=\frac{7}{2}=3\frac{1}{2}$ yd

e. ex. large $=2\frac{1}{8}$ yd, large $=1\frac{3}{4}$ yd

$2\frac{1}{8}-1\frac{3}{4}=\frac{17}{8}-\frac{7}{4}=\frac{17}{8}-\frac{14}{8}=\frac{3}{8}$ yd

f. $21\text{ yd}\div1\frac{3}{4}\text{ yd}=21\div\frac{7}{4}=21\cdot\frac{4}{7}$

$=\frac{84}{7}=12$ shirts

g. $1\frac{1}{2}+2\frac{1}{8}=\frac{3}{2}+\frac{17}{8}=\frac{12}{8}+\frac{17}{8}=\frac{29}{8}=3\frac{5}{8}$ yd

h. $2\frac{1}{8}-1\frac{1}{2}=\frac{17}{8}-\frac{3}{2}=\frac{17}{8}-\frac{12}{8}=\frac{5}{8}$ yd

39.

$1\quad 2\quad 3\quad 4$

Dots	Line Segments
1	2
2	3
3	4
4	5
10	11

41.

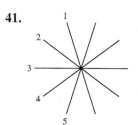

Lines	Regions
1	2
2	4
3	6
4	8
5	10
10	20

43.

Credit Hours	Tuition Cost (dollars)
8	$8(55)=440$
9	$9(55)=495$
10	$10(55)=550$
11	600
12	600

45. Answers may vary.

47. a. $150 is between $0 and $166.66 so the fee is $5.

2

b.　$300 is between $166.67 and $1333.33 so the fee is 3% of $300.

$$3\% = \frac{3}{100} = 0.03$$

$$0.03(\$300) = \$9$$

The fee is $9.

c.　$1000 is between $166.67 and $1333.33 so the fee is 3% of $1000.

$$3\% = \frac{3}{100} = 0.03$$

$$0.03(\$1000) = \$30$$

The fee is $30.

d.　$2050 is above $1333.34 so the fee is $40.

e.　The cash advance equals the fee.

f.　$0.03(\$166.67) = \$5.0001$

$0.03(\$1333.33) = \$39.9999$

These are the amounts for which $5 equals 3% and $40 equals 3%.

49.

a	b	$a + b$	$a - b$
15	5	$15 + 5 = 20$	$15 - 5 = 10$
$\frac{3}{4}$	$\frac{2}{5}$	$\frac{3}{4} + \frac{2}{5} = \frac{15+8}{20} = 1\frac{3}{20}$	$\frac{3}{4} - \frac{2}{5} = \frac{15-8}{20} = \frac{7}{20}$
0.36	0.06	$0.36 + 0.06 = 0.42$	$0.36 - 0.06 = 0.30$
5.6	0.7	$5.6 + 0.7 = 6.3$	$5.6 - 0.7 = 4.9$
2.25	1.5	$2.25 + 1.5 = 3.75$	$2.25 - 1.5 = 0.75$

a	b	$a \cdot b$	$a \div b$
15	5	$15 \cdot 5 = 75$	$15 \div 5 = 3$
$\frac{3}{4}$	$\frac{2}{5}$	$\frac{3}{4} \cdot \frac{2}{5} = \frac{6}{20} = \frac{3}{10}$	$\frac{3}{4} \div \frac{2}{5} = \frac{3}{4} \cdot \frac{5}{2} = \frac{15}{8} = 1\frac{7}{8}$
0.36	0.06	$0.36 \cdot 0.06 = 0.0216$	$0.36 \div 0.06 = 6$
5.6	0.7	$5.6 \cdot 0.7 = 3.92$	$5.6 \div 0.7 = 8$
2.25	1.5	$2.25 \cdot 1.5 = 3.375$	$2.25 \div 1.5 = 1.5$

51.

a	b
$15 = \dfrac{15 \cdot 100}{100} = \dfrac{1500}{100} = 1500\%$	$5 = \dfrac{5 \cdot 100}{100} = \dfrac{500}{100} = 500\%$
$\dfrac{3}{4} = \dfrac{3 \cdot 25}{4 \cdot 25} = \dfrac{75}{100} = 75\%$	$\dfrac{2}{5} = \dfrac{2 \cdot 20}{5 \cdot 20} = \dfrac{40}{100} = 40\%$
$0.36 = \dfrac{36}{100} = 36\%$	$0.06 = \dfrac{6}{100} = 6\%$
$5.6 = \dfrac{5.6 \cdot 100}{100} = \dfrac{560}{100} = 560\%$	$0.7 = \dfrac{7}{10} = \dfrac{7 \cdot 10}{10 \cdot 10} = \dfrac{70}{100} = 70\%$
$2.25 = \dfrac{2.25 \cdot 100}{100} = \dfrac{225}{100} = 225\%$	$1.5 = \dfrac{1.5 \cdot 100}{100} = \dfrac{150}{100} = 150\%$

Mid-Chapter 1 Test

1.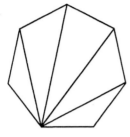

Sides	Triangles
3	1
4	2
5	3
6	4
7	5
20	18

2. **a.** True

   b. True

   c. False. Zero is added to the set of natural numbers to make the set of whole numbers.

   d. False. 3 divided by 4 is not an integer.

3. **a.** $2\frac{1}{3}+3\frac{3}{4}=\frac{7}{3}+\frac{15}{4}=\frac{28}{12}+\frac{45}{12}=\frac{73}{12}=6\frac{1}{12}$

   b. $1\frac{1}{2}\cdot1\frac{1}{4}=\frac{3}{2}\cdot\frac{5}{4}=\frac{15}{8}=1\frac{7}{8}$

   c. $3\frac{1}{4}-1\frac{3}{4}=\frac{13}{4}-\frac{7}{4}=\frac{6}{4}=1\frac{1}{2}$

   d. $3\frac{1}{2}\div1\frac{3}{4}=\frac{7}{2}\div\frac{7}{4}=\frac{7}{2}\cdot\frac{4}{7}=2$

4. **a.** $3\cdot(\$0.35)=\$1.05$

   b. $4\cdot(\$0.295)=\$1.18$

   c. $6\cdot(\$0.185)=\$1.11$

   d. $\dfrac{\$2.45}{\$0.35}=7$
   We could talk for 7 minutes.

Exercises 1.3

1. c: 3 times the input, plus 1

Input	Output
4	$3(4)+1=13$
50	$3(50)+1=151$
100	$3(100)+1=301$
n	$3n+1$

3. b: 1 less than 4 times the input

Input	Output
4	$4(4)-1=15$
50	$4(50)-1=199$
100	$4(100)-1=399$
n	$4n-1$

5. **a.** $2\pi r$; 2 and π are both constants and numerical coefficients, r is a variable.

   b. $1.5x$; 1.5 is a constant and a numerical coefficient, and x is a variable.

   c. $-4n+3$; -4 is a constant and a numerical coefficient, 3 is a constant, and n is a variable.

   d. x^2-9 ; The implied 1 (before the x^2) is a constant and numerical coefficient, -9 is a constant, and x is a variable.

7. Answers may vary.

9. Answers may vary.

11. **a.** 35% of $n=\dfrac{35}{100}\cdot n=0.35n$

   b. 10% of $x=\dfrac{10}{100}\cdot x=0.10x$

   c. $87\frac{1}{2}$% of $n=\dfrac{87.5}{100}\cdot n=0.875n$

   d. $37\frac{1}{2}$% of $x=\dfrac{37.5}{100}\cdot x=0.375x$

   e. $\frac{1}{2}$% of $n=\dfrac{0.5}{100}\cdot n=0.005n$

   f. 108% of $x=\dfrac{108}{100}\cdot x=1.08x$

13. **a.** $3\cdot n$ or $3n$

   b. $8\div n$ or $\dfrac{8}{n}$

   c. $n-4$

   d. $n\div5$ or $\dfrac{n}{5}$

    e. $15 \cdot \$n = \$15n$

15. a. $3 + 2 \cdot n = 3 + 2n$

    b. $4 - 3 \cdot n = 4 - 3n$

    c. $7 \cdot n + 4 = 7n + 4$

    d. $n \cdot n = n^2$

    e. $n \cdot \$0.79 = \$0.79n$

17. Not necessarily. x is a variable and could represent a number that is positive, negative, or even zero.

19.

Input	Output
1	$55(1) = 55$
2	$55(2) = 110$
3	$55(3) = 165$
t	$55 \cdot t = 55t$

21.

Input	Output
1	$\$250 + \$75(1) = \$325$
2	$\$250 + \$75(2) = \$400$
3	$\$250 + \$75(3) = \$475$
n	$\$250 + \$75(n) = \$250 + \$75n$

23. Let $x =$ input ; ouput $= 4x + 2$

Input	Output
1	$4(1) + 2 = 6$
2	$4(2) + 2 = 10$
3	$4(3) + 2 = 14$
4	$4(4) + 2 = 18$
5	$4(5) + 2 = 22$

25.

Input, n	Output Rule
$0 to $166.66	$5
$166.67 to $1333.33	$0.03n$
$1333.34 to limit	$40

27. Rule: output = 2 more than twice the input

Tables	Chairs
4	$2(4) + 2 = 10$
n	$2(n) + 2 = 2n + 2$
20	$2(20) + 2 = 42$

29. Rule: output = 5 if input is even and twice the input if the input is odd.

Input	Output
0	5
1	$2(1) = 2$
2	5
3	$2(3) = 6$
4	5
5	$2(5) = 10$
6	5
7	$2(7) = 14$
8	5

31.

Input	Output
8	9
50	51
101	202
Even n	$n + 1$
Odd n	$2n$

33.

Input x	Input y	Output xy	Output $x + y$
4	3	12	7
1	12	12	13
5	4	20	9
20	1	20	21
2	2	4	4
6	3	18	9

35. a. Initial guesses may vary.
    $3 + 9 = 12$ and $3 \cdot 9 = 27$
    The numbers are 3 and 9.

    b. Initial guesses may vary.
    $3 + 7 = 10$ and $3 \cdot 7 = 21$
    The numbers are 3 and 7.

Exercises 1.4

1. $A(-2, 2)$, $B(-5, 0)$, $C(-3, -4)$, $D(0, -2)$,
    $E(4, -4)$, $F(2, -6)$,
    $G(4, 0)$, $H(4, 6)$, $I(0, 5)$

3. **a.** Quadrant 2 (x is negative and y is positive)

   b. Quadrant 4 (x is positive and y is negative)

   c. Quadrant 3 (both x and y are negative)

   d. Quadrant 3 (both x and y are negative)

5. **a.** vertical (x = 0)

   b. vertical (x = 0)

   c. horizontal (y = 0)

7.

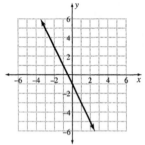

9.

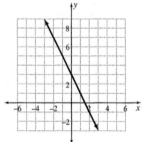

11. – 17.

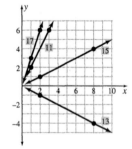

19.

Input, x	Output, $y = 2x + 5$
0	$2(0) + 5 = 5$
1	$2(1) + 5 = 7$
2	$2(2) + 5 = 9$
3	$2(3) + 5 = 11$
4	$2(4) + 5 = 13$
5	$2(5) + 5 = 15$

21.

Input, x	Output, $y = \dfrac{1}{2}x$
0	$\dfrac{1}{2}(0) = 0$
1	$\dfrac{1}{2}(1) = \dfrac{1}{2}$ or 0.5
2	$\dfrac{1}{2}(2) = 1$
3	$\dfrac{1}{2}(3) = \dfrac{3}{2}$ or 1.5
4	$\dfrac{1}{2}(4) = 2$
5	$\dfrac{1}{2}(5) = \dfrac{5}{2}$ or 2.5

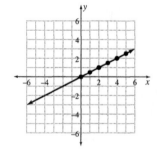

6

23.

Input, x	Output, $y = 3x - 2$
0	$3(0) - 2 = -2$
1	$3(1) - 2 = 1$
2	$3(2) - 2 = 4$
3	$3(3) - 2 = 7$
4	$3(4) - 2 = 10$
5	$3(5) - 2 = 13$

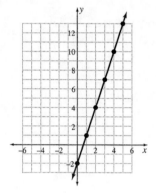

25.

Input, x	Output, $y = 8 - x$
0	$8 - 0 = 8$
1	$8 - 1 = 7$
2	$8 - 2 = 6$
3	$8 - 3 = 5$
4	$8 - 4 = 4$
5	$8 - 5 = 3$

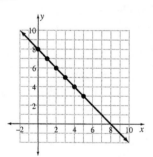

27. a.

Input x lbs.	Output y dollars
0	$6.50(0) = 0$
1	$6.50(1) = 6.50$
2	$6.50(2) = 13.00$
3	$6.50(3) = 19.50$
4	$6.50(4) = 26.00$

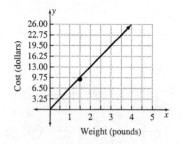

b. From the graph we estimate the cost to be about \$16. Using the price per pound we get: \$6.50(2.5) = \$16.25.

From the graph we estimate the cost to be about \$11. Using the price per pound we get: \$6.50(1.75) = \$11.38.

From the graph we estimate the cost to be about \$21. Using the price per pound we get: \$6.50(3.25) = \$21.13

c. Packaged nuts are a better deal, the data point is below the line on the graph. 1.5 pound of bulk nuts would cost \$9.75

29. a.

Input Minutes	Value Remaining (\$)
0	$24 - 1.50(0) = 24.00$
4	$24 - 1.50(4) = 18.00$
8	$24 - 1.50(8) = 12.00$
12	$24 - 1.50(12) = 6.00$
16	$24 - 1.50(16) = 0$

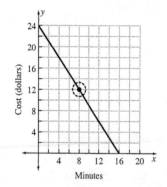

b. \$12.00 remains

c. In (0, 24), 24 is the value of the card after talking zero minutes.

d. In (16, 0), 16 is the total minutes of phone time used on the card at \$1.50 per minute. After 16 minutes, the value of the card is zero.

31. a.

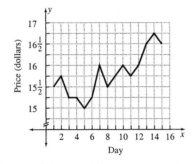

b. Prices are rising slightly.

33. The order in which the numbers are written makes a difference.

35. From the ordered pair (x, y), trace vertically to the horizontal axis to find x and trace horizontally to the vertical axis to find y.

37. Points on the horizontal axis have ordered pairs whose second coordinate is zero. Points on the vertical axis have ordered pairs whose first coordinate is zero.

39. a. A is 2 left of and 2 below $(2,0)$

change in $x = -2$, change in $y = -2$

$2+(-2) = 0$, $0+(-2) = -2$

$A = (0,-2)$

B is 5 left of $(2,0)$

change in $x = -5$, change in $y = 0$

$2+(-5) = -3$, $0+0 = 0$

$B = (-3,0)$

b. A is 2 right of and 3 below $(0,-2)$

change in $x = 2$, change in $y = -3$

$0+2 = 2$, $-2+(-3) = -5$

$A = (2, \ -5)$

B is 2 left of and 3 below $(0,-2)$

change in $x = -2$, change in $y = -3$

$0+(-2) = -2$, $-2+(-3) = -5$

$B = (-2, \ -5)$

c. A is 2 right of and 1 above $(4,0)$

change in $x = 2$, change in $y = 1$

$4+2 = 6$, $0+1 = 1$

$A = (6,1)$

B is 2 below $(4,0)$

change in $x = 0$, change in $y = -2$

$4+0 = 4$, $0+(-2) = -2$

$B = (4,-2)$

d. A is 1 right of and 1 above $(2, 3)$

change in $x = 1$, change in $y = 1$

$2 + 1 = 3$, $3 + 1 = 4$

$A = (3, 4)$

B is 3 right of and 1 below $(2, 3)$

change in $x = 3$, change in $y = -1$

$2+3 = 5$, $3+(-1) = 2$

$B = (5, 2)$

41.

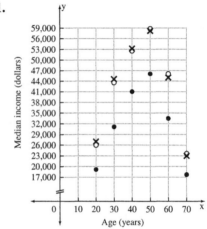

Ages 50 and older lost income. This is possibly due to a weak stock market.

Chapter 1 Review Exercises

1. a. You have time after class to reach the bus stop before 3:15.

b. Write the 8 first and subtract 5.

c. Write the 12 first and divide by 8.

d. Your bicycle is in good working condition.

3.

Pens	Panels
4	$3(4)+1 = 13$
n	$3n+1$

Rule: output equals one more than three times the input. ($3n+1$)

$n = 20: \quad 3(20) + 1 = 61$

$n = 50: \quad 3(50) + 1 = 151$

20 pens will require 61 panels and 50 pens will require 151 panels.

5. a. Denominator

   b. Numerator

   c. Numerator

   d. Denominator

   e. Numerator

   f. Denominator

   g. Denominator

7. a. $\quad 2\frac{3}{4} + 1\frac{1}{2} + 3\frac{1}{4} = 6 + 1\frac{1}{2} = 7\frac{1}{2}$

Add the terms with the same denominator first. Since $\frac{3}{4} + \frac{1}{4} = \frac{4}{4} = 1$, the result is a whole number which we can add to the remaining term.

   b. $\quad 2\frac{1}{2} + 3\frac{1}{3} + 4\frac{2}{3} = 2\frac{1}{2} + 8 = 10\frac{1}{2}$

Add the terms with the same denominator first. Since $\frac{1}{3} + \frac{2}{3} = \frac{3}{3} = 1$, the result is a whole number which we can add to the remaining term.

   c. $\quad \$1.75 + \$2.98 + \$3.25$

$= \$5.00 + \$2.98$

$= \$7.98$

Add the first and last terms together first. Since $0.75 + 0.25 = 1.00$, the result is a whole number which we can add to the remaining term.

   d. $\quad \$6.99 + \$4.20 + \$2.80$

$= \$6.99 + \$7.00$

$= \$13.99$

Add the last two terms together first. Since $0.20 + 0.80 = 1.00$, the result is a whole number which we can add to the remaining term.

9. a. $\quad \dfrac{5}{6} + \dfrac{3}{8} = \dfrac{20}{24} + \dfrac{9}{24} = \dfrac{29}{24} = 1\dfrac{5}{24}$

$\dfrac{5}{6} - \dfrac{3}{8} = \dfrac{20}{24} - \dfrac{9}{24} = \dfrac{11}{24}$

$\dfrac{5}{6} \cdot \dfrac{3}{8} = \dfrac{15}{48} = \dfrac{5}{16}$

$\dfrac{5}{6} \div \dfrac{3}{8} = \dfrac{5}{6} \cdot \dfrac{8}{3} = \dfrac{40}{18} = \dfrac{20}{9} = 2\dfrac{2}{9}$

   b. $\quad 2\dfrac{1}{6} + 1\dfrac{3}{4} = \dfrac{13}{6} + \dfrac{7}{4} = \dfrac{26}{12} + \dfrac{21}{12} = \dfrac{47}{12} = 3\dfrac{11}{12}$

$2\dfrac{1}{6} - 1\dfrac{3}{4} = \dfrac{26}{12} - \dfrac{21}{12} = \dfrac{5}{12}$

$2\dfrac{1}{6} \cdot 1\dfrac{3}{4} = \dfrac{13}{6} \cdot \dfrac{7}{4} = \dfrac{91}{24} = 3\dfrac{19}{24}$

$2\dfrac{1}{6} \div 1\dfrac{3}{4} = \dfrac{13}{6} \div \dfrac{7}{4} = \dfrac{13}{6} \cdot \dfrac{4}{7} = \dfrac{26}{21} = 1\dfrac{5}{21}$

11. a. real, rational

    b. real, rational, integer, whole number

    c. real, rational

    d. real, rational

    e. real, rational

    f. real, rational, integer

13. The constant term is 3.

15. a. Four less than three times the input.

    b. Three more than the input times itself.

    c. The input divided by three.

17.

Kilowatt Hours	Cost (dollars)
10	$5.50 + 0.10(10) = 6.50$
20	$5.50 + 0.10(20) = 7.50$
50	$5.50 + 0.10(50) = 10.50$
h	$5.50 + 0.10h$

19. a. The input is the figure number (or the number of circles in the middle column of the figure). The output is the total number of circles.

b.

Input	Output
50	54
100	104

c. The output is four more than the input.
$n + 4$

21.

Input n	Output $n \div 2$
0	$0 \div 2 = 0$
1	$1 \div 2 = 0.5$
2	$2 \div 2 = 1$
3	$3 \div 2 = 1.5$
4	$4 \div 2 = 2$
5	$5 \div 2 = 2.5$
6	$6 \div 2 = 3$

23.

Input	Output
0	$2(0) = 0$
1	4
2	$2(2) = 4$
3	4
4	$2(4) = 8$
5	4
6	$2(6) = 12$

25. a. $65 is between $0 and $149.99 so there is a 5% discount.
$0.05(65) = 3.25$
The discount is $3.25.

b. $145 is between $0 and $149.99 so there is a 5% discount.
$0.05(145) = 7.25$
The discount is $7.25.

c. $250 is between $150 and $499.99 so there is a 6% discount.
$0.06(250) = 15$
The discount is $15.00.

d. $500 receives a discount of $50 or 8%, whichever is greater.

$0.08(500) = 40$
The discount is $50.00.

e. $550 is over $500 so the discount is $50 or 8%, whichever is greater.
$0.08(550) = 44$
The discount is $50.00.

f. $700 is over $500 so the discount is $50 or 8%, whichever is greater.
$0.08(750) = 60$
The discount is $60.00.

27. a.-j.

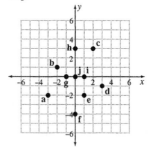

29.

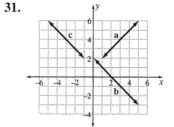

31.

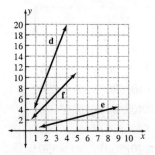

(answer in book shows points connected, but the directions do not indicate to do so)

33. Let x be the input and y be the output.

   a. $y = x + 1$

   b. $y = -x + 2$

   c. $y = -x + 1$

   d. $y = 5x$

   e. $y = \dfrac{x}{2}$

   f. $y = 2x + 1$

35.

Input (pounds)	Output (dollars)
0	$24(0) = 0$
1	$24(1) = 24$
2	$24(2) = 48$
3	$24(3) = 72$
4	$24(4) = 96$
5	$24(5) = 120$
6	$24(6) = 144$
7	$24(7) = 168$
8	$24(8) = 192$

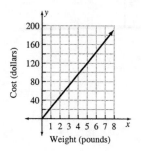

37. a. $\dfrac{\$9.99}{30} = \dfrac{(9.99)(100)}{30}$ cents

          $= \dfrac{999}{30}$ cents

          $= 33\dfrac{1}{3}$ cents ≈ 33.3 cents

    Each minute costs about 33.3 cents.

  b.

Input (min)	Output: Value (dollars)
0	$9.99 - 0.333(0) = 9.99$
5	$9.99 - 0.333(5) = 8.325$
10	$9.99 - 0.333(10) = 6.66$
15	$9.99 - 0.333(15) = 4.995$
20	$9.99 - 0.333(20) = 3.33$
25	$9.99 - 0.333(25) = 1.665$
30	$9.99 - 0.333(30) = 0$

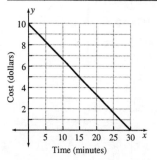

39. Inputs can be any positive number.

41. a.   Answers may vary.
        Weight loss starts off slow but the rate at which weight is lost increases as time goes on.

  b.   Answers may vary.
        Weight loss starts off quickly but the rate at which weight is lost decreases as time goes on.

  c.   Answers may vary.
        The rate of weight loss is constant.

Chapter 1 Test

1. Answers may vary. Some possible answers: Material on the test will be similar to the chapter exercises and chapter review; the test will cover the entire chapter; the number of questions from each section will be about the same.

2. Answers may vary. Some possible answers:
   The test has a time limit; students must work
   on their own; the test has the same weight
   towards the final grade as other chapter tests.

3. difference

4. set

5. integers

6. Label *a* identifies the origin, label *b* identifies
   the vertical (or y) axis, label *c* identifies the
   horizontal (or x) axis, and label *d* identifies
   Quadrant 3.

7. $e(-4,4)$, $f(2,-3)$, $g(0,-5)$, and $h(-2,-7)$

8.

Input x	Output $y = 2x + 3$
0	$2(0) + 3 = 3$
1	$2(1) + 3 = 5$
2	$2(2) + 3 = 7$
3	$2(3) + 3 = 9$
4	$2(4) + 3 = 11$
5	$2(5) + 3 = 13$

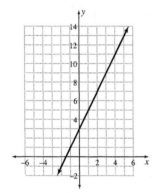

9. **a.** Miles go on the *x*-axis and total cost (in
   dollars) goes on the *y*-axis.
   An expression for the rental would be
   $\$0.40x + \$150$

   b. Answers may vary.
   Consider that the minimum cost is $150
   and that the number of miles goes from 0
   to 600 by 100.

c.

Input (miles)	Output: Cost (dollars)
0	$0.40(0) + 150 = 150$
100	$0.40(100) + 150 = 190$
200	$0.40(200) + 150 = 230$
300	$0.40(300) + 150 = 270$
400	$0.40(400) + 150 = 310$
500	$0.40(500) + 150 = 350$
600	$0.40(600) + 150 = 390$

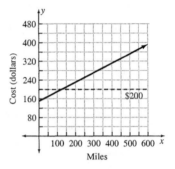

d. Locate 200 on the vertical axis, move
   horizontally to the graphed line, then
   move vertically down to the horizontal
   axis to determine the number of miles.

   The second car becomes a better deal if
   you drive more than 125 miles.

10.

Input	Output
0	$0 \div 2 = 0$
1	$2(1) = 2$
2	$2 \div 2 = 1$
3	$2(3) = 6$
4	$4 \div 2 = 2$
5	$2(5) = 10$
6	$6 \div 2 = 3$
7	$2(7) = 14$
8	$8 \div 2 = 4$

11. $\dfrac{5}{8} + \dfrac{1}{6} = \dfrac{15}{24} + \dfrac{4}{24} = \dfrac{19}{24}$

12. $1\dfrac{1}{9} \cdot 1\dfrac{5}{12} = \dfrac{10}{9} \cdot \dfrac{17}{12}$

$\qquad\qquad = \dfrac{\cancel{2} \cdot 5 \cdot 17}{9 \cdot \cancel{2} \cdot 6}$

$\qquad\qquad = \dfrac{85}{54}$

$\qquad\qquad = 1\dfrac{31}{54}$

13. 9% of $35 = 0.09(35) = 3.15$

14.

x	y
5	20
100	400
n	$4n$

The output is four times the input.

15. A is 4 left of and 2 above $(-1, 2)$

change in $x = -4$, change in $y = 2$

$-1 + (-4) = -5$, $2 + 2 = 4$

$A = (-5, 4)$

B is 1 left of and 3 above $(-1, 2)$

change in $x = -1$, change in $y = 3$

$-1 + (-1) = -2$, $2 + 3 = 5$

$B = (-2, 5)$

Chapter 2

Exercises 2.1

1. a. -5

   b. $\dfrac{1}{2}$

   c. -0.4

   d. $-x$

   e. $2x$

3. a. $|4| = 4$

   b. $|-6| = 6$

   c. $-(-5) = -1 \cdot (-5) = 5$

   d. $-(-2) = -1 \cdot (-2) = 2$

5. a. $-|7| = -1 \cdot |7| = -1 \cdot 7 = -7$

   b. $-|-8| = -1 \cdot |-8| = -1 \cdot 8 = -8$

   c. $-(-3) = -1 \cdot (-3) = 3$

   d. $|-7| = 7$

7. a. $|-4| = 4$

   b. $|5| = 5$

   c. $|4-9| = |-5| = 5$

   d. $-|2+5| = -|7| = -1 \cdot 7 = -7$

9. a. $-(-4) = -1 \cdot (-4) = 4$

   b. $-(-(-4)) = -1 \cdot (-1 \cdot (-4))$
      $= -1 \cdot 4 = -4$

11. a. $-|6| = -1 \cdot 6 = -6$

    b. $-|-6| = -1 \cdot |-6| = -1 \cdot 6 = -6$

13. The opposite of the absolute value of x.

15. $+4 + (-3) = 1$
    $-3 + (+4) = 1$
    net charge $= 1$

17. $+5 + (-5) = 0$
    $-5 + (+5) = 0$
    net charge $= 0$

19. $+5 + (-7) = -2$
    $-7 + (+5) = -2$
    net charge $= -2$

21. a. $-8 + 3 = -5$

    b. $4 + (-7) = -3$

    c. $+4 + (-4) = 0$

23. a. $-3 + (+3) = 0$

    b. $-4 + (-7) = -11$

    c. $-12 + (+8) = -4$

25. a. $-3 + (+2) = -1$

    b. $+6 + (-4) = 2$

    c. $-14 + (-6) = -20$

27. a. $-8 + (-17) = -25$

    b. $-9 + (+16) = 7$

    c. $24 + (-8) = 16$

29. a. $+1 - (-2) = +1 + (+2) = 3$

    b. $+2 - (+3) = +2 + (-3) = -1$

31. a. $0 - (+2) = 0 + (-2) = -2$

    b. $+1 - (-3) = +1 + (+3) = 4$

33. a. $8 - 11 = 8 + (-11) = -3$

   b. $-8-(-11)=-8+11=3$

   c. $-16-3=-16+(-3)=-19$

35. a. $12-7=12+(-7)=5$

   b. $-12-7=-12+(-7)=-19$

   c. $-17-(-4)=-17+(+4)=-13$

37. a. $-4-(-4)=-4+(4)=0$

   b. $14-(10)=14+(-10)=4$

   c. $0-(-5)=0+(+5)=5$

39. a. $10-14=10+(-14)=-4$

   b. $-8-(-9)=-8+(+9)=1$

   c. $-7-6=-7+(-6)=-13$

41. a. $2-7=2+(-7)=-5$

   b. $2-5=2+(-5)=-3$

   c. $2-(-5)=2+(+5)=7$

43. a. $-7-5=-7+(-5)=-12$

   b. $-5-7=-5+(-7)=-12$

   c. $-5-(-7)=-5+(+7)=2$

45. a. $6960-(-40)=6960+40=7000$ m
     The difference in height is 7000 m.

   b. $5896-(-156)=5896+156=6052$ m
     The difference in height is 6052 m.

   c. $2228-(-16)=2228+16=2244$ m
     The difference in height is 2244 m.

   d. $-10,930-(-8600)=-10,930+8600$
$$=-2330$$
     The Marianna Trench is 2330 m deeper
     than the Puerto Rico trench.

47. True. They are the same distance away from 0
    on a number line.

49. The absolute value of a number is the distance
    it is from 0 on a number line.

51. To subtract, change to adding the opposite.
    That is, $a-b=a+(-b)$.

53. With each negative subtracted, a positive
    charge is left unmatched.

Exercises 2.2

  1. $-2(+150)=-300$
     The net change is $-\$300$.

  3. $+2(+400)=800$
     The net change is $\$800$.

  5. $-8(-40)=320$
     The net change is $\$320$.

  7. $+3(-90)=-270$
     The net change is $-\$270$.

  9.

Input x	Output $5x$
2	$5(2)=10$
1	$5(1)=5$
0	$5(0)=0$
−1	$5(-1)=-5$
−2	$5(-2)=-10$

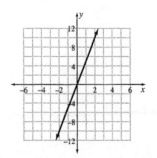

11.

Input x	Output $-2x$
2	$-2(2) = -4$
1	$-2(1) = -2$
0	$-2(0) = 0$
-1	$-2(-1) = 2$
-2	$-2(-2) = 4$

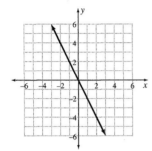

13. a. +, same sign: $-7(-6) = 42$

  b. −, opposite signs: $6(-8) = -48$

  c. −, opposite signs: $3(-15) = -45$

  d. +, same sign: $-5(-7) = 35$

15. a. −, opposite signs: $12(-4) = -48$

  b. −, opposite signs: $8(-8) = -64$

  c. +, same sign: $15(4) = 60$

  d. +, same sign: $-9(-9) = 81$

17. a. −, opposite signs: $-3\left(\dfrac{1}{3}\right) = -\dfrac{3}{3} = -1$

  b. +, same sign: $-3\left(-\dfrac{1}{3}\right) = \dfrac{3}{3} = 1$

  c. −, opposite signs: $2\left(-\dfrac{1}{2}\right) = -\dfrac{2}{2} = -1$

  d. +, same sign: $-\dfrac{1}{4}(-4) = \dfrac{4}{4} = 1$

19. a. −, opposite signs: $\dfrac{3}{4}(-8) = -\dfrac{24}{4} = -6$

  b. −, opposite signs: $-\dfrac{1}{5}(5) = -\dfrac{5}{5} = -1$

  c. +, same sign: $-4\left(-\dfrac{3}{4}\right) = \dfrac{12}{4} = 3$

21. a. +, same sign: $-\dfrac{5}{3}\left(-\dfrac{6}{7}\right) = \dfrac{30}{21} = \dfrac{10}{7}$

  b. −, opposite signs:
  $-\dfrac{1}{6}(9) = -\dfrac{9}{6} = -\dfrac{3}{2}$

  c. $-\dfrac{2}{3}(0) = 0$

23. a. $\dfrac{1}{4}$

  b. $-\dfrac{1}{2}$

  c. $\dfrac{2}{1} = 2$

  d. $-\dfrac{4}{3}$

  e. $\dfrac{1}{0.5} = \dfrac{1}{1/2} = \dfrac{2}{1} = 2$

25. a. $3\dfrac{1}{3} = \dfrac{10}{3}$; reciprocal $= \dfrac{3}{10}$

  b. $6.5 = 6\dfrac{1}{2} = \dfrac{13}{2}$; reciprocal $= \dfrac{2}{13}$

  c. $\dfrac{1}{x}$

  d. $\dfrac{b}{a}$

  e. $-\dfrac{1}{x}$

27. a. $15 \div (-5) = 15\left(-\dfrac{1}{5}\right) = -\dfrac{15}{5} = -3$

b. $-45 \div 9 = -45\left(\dfrac{1}{9}\right) = -\dfrac{45}{9} = -5$

c. $42 \div (-6) = 42\left(-\dfrac{1}{6}\right) = -\dfrac{42}{6} = -7$

29. a. $\dfrac{+}{-} = -$; $\dfrac{56}{-8} = -7$

b. $\dfrac{-}{-} = +$; $\dfrac{-56}{-8} = 7$

c. $- \cdot + = -$; $-\dfrac{56}{8} = -7$

31. a. $\dfrac{-}{-} = +$; $\dfrac{-55}{-11} = 5$

b. $\dfrac{-}{+} = -$; $\dfrac{-28}{4} = -7$

c. $\dfrac{-}{+} = -$; $\dfrac{-27}{9} = -3$

33. a. $+ \div + = +$; $24 \div \dfrac{2}{3} = 24\left(\dfrac{3}{2}\right) = \dfrac{72}{2} = 36$

b. $+ \div - = -$; $24 \div \left(-\dfrac{3}{2}\right) = 24\left(-\dfrac{2}{3}\right)$

$\qquad\qquad = -\dfrac{48}{3}$

$\qquad\qquad = -16$

c. $- \div + = -$; $-16 \div \dfrac{3}{8} = -16 \cdot \dfrac{8}{3} = -\dfrac{128}{3}$

35. a. $- \div + = -$; $-16 \div \dfrac{8}{5} = -16 \cdot \dfrac{5}{8} = -\dfrac{80}{8} = -10$

b. $+ \div - = -$; $24 \div \left(-\dfrac{3}{4}\right) = 24 \cdot \left(-\dfrac{4}{3}\right)$

$\qquad\qquad = -\dfrac{96}{3}$

$\qquad\qquad = -32$

c. $- \div - = +$;

$\qquad -\dfrac{3}{4} \div \left(-\dfrac{6}{7}\right) = -\dfrac{3}{4}\left(-\dfrac{7}{6}\right) = \dfrac{21}{24} = \dfrac{7}{8}$

37.

a	b	$-\left(\dfrac{b}{a}\right)$	$\dfrac{-b}{a}$	$\dfrac{b}{-a}$
5	35	$-\left(\dfrac{35}{5}\right)$ $= -7$	$\dfrac{-35}{5}$ $= -7$	$\dfrac{35}{-5}$ $= -7$
-27	3	$-\left(\dfrac{3}{-27}\right)$ $= \dfrac{1}{9}$	$\dfrac{-3}{-27}$ $= \dfrac{1}{9}$	$\dfrac{3}{-(-27)}$ $= \dfrac{1}{9}$

39.

x	y	$x \cdot y$	$x + y$
2	-2	$2(-2) = -4$	$2 + (-2) = 0$
3	$-6 \div 3 = -2$	-6	$3 + (-2) = 1$
-4	$-12 \div -4 = 3$	-12	$-4 + 3 = -1$
$0 - 3 = -3$	3	$-3(3) = -9$	0
2	$-1 - 2 = -3$	$2(-3) = -6$	-1

41.

Input	Output
-3	$-1(-3) = 3$
-2	$-1(-2) = 2$
-1	$-1(-1) = 1$
0	0
1	1
2	2
3	3

43. $-\dfrac{b}{a}$ and $\dfrac{-b}{a}$

45. Answers may vary. Some possible answers:

Multiply the numbers and write the answer as a positive.

Multiply the absolute values of the numbers.

47. $\dfrac{\frac{1}{2}}{\frac{2}{3}} = 1 \div \dfrac{2}{3} = 1 \cdot \dfrac{3}{2} = \dfrac{3}{2}$

The answers are the same.

Exercises 2.3

1. **a.** Associative property of multiplication

   b. Associative property of addition

   c. Associative property of addition

   d. Commutative properties

   e. Associative property of addition

   f. Associative property of multiplication

   g. Distributive property of multiplication over division

3. **a.** $2\frac{1}{2}+1\frac{1}{3}+3\frac{2}{3}+5\frac{1}{2}$

   $=(2+1+3+5)+\left(\frac{1}{2}+\frac{1}{2}\right)+\left(\frac{1}{3}+\frac{2}{3}\right)$

   $=11+1+1$

   $=13$

   b. $1\frac{2}{5}+3\frac{1}{3}+4\frac{3}{5}$

   $=(1+3+4)+\left(\frac{2}{5}+\frac{3}{5}\right)+\frac{1}{3}$

   $=8+1+\frac{1}{3}$

   $=9\frac{1}{3}$

5. **a.** $4.25+2.98+1.75m=(4.25+1.75)+2.98$

   $=6+2.98$

   $=8.98$

   b. $2.60+1.30+3.40=(2.60+3.40)+1.30$

   $=6+1.30$

   $=7.30$

7. **a.** $-3+(-4)+(+5)=-7+(+5)=-2$

   b. $+7+(-8)+(-3)=+7+(-11)=-4$

9. **a.** $2+(-5)+(-4)=2+(-9)=-7$

   b. $-7+(-3)+(+7)=(-3)+(-7)+(+7)$

   $=-3+0$

   $=-3$

11. $-12+(+6)+(-15)+(+7)$

    $=(-12)+(-15)+(+6)+(+7)$

    $=(-27)+(+13)$

    $=-14$

13. $-6+(-7)+(-8)+20=-13+(-8)+20$

    $=-21+20$

    $=-1$

15. **a.** $-4-(-3)=-4+(+3)=-1$

    b. $-1-(-3)=-1+(+3)=2$

17. **a.** $8-(+3)-(-4)=8+(-3)+(+4)$

    $=5+(+4)$

    $=9$

    b. $-5-(-7)+(-1)=-5+(+7)+(-1)$

    $=2+(-1)$

    $=1$

19. **a.** $\frac{1}{4}\cdot25\cdot8\cdot4=25\cdot8\cdot4\cdot\frac{1}{4}$

    $=25\cdot8\cdot1$

    $=25\cdot8$

    $=200$

    b. $8\left(\frac{2}{3}\right)(1.5)=8\left(\frac{2}{3}\right)\left(1\frac{1}{2}\right)$

    $=8\left(\frac{2}{3}\right)\left(\frac{3}{2}\right)$

    $=8\cdot1$

    $=8$

21. **a.** $6(-8)(-1)=6(8)=48$

    b. $-5(5)(-1)=-5(-5)=25$

    c. $4(3)(-2)=12(-2)=-24$

23. **a.** $-5(-6)(-7)=30(-7)=-210$

    b. $-6(9)(2)=-6(18)=-108$

    c. $-3(0)(-7)=0(-7)=0$

18

25. $4 \cdot \$4.97 = 4(\$5.00 - \$0.03)$
$\qquad = 4 \cdot \$5.00 - 4 \cdot \$0.03$
$\qquad = \$20.00 - \$0.12$
$\qquad = \$19.88$

27. $3 \cdot \$10.98 = 3(\$11.00 - \$0.02)$
$\qquad = 3 \cdot \$11.00 - 3 \cdot \$0.02$
$\qquad = \$33.00 - \$0.06$
$\qquad = \$32.94$

29. $4\left(2\dfrac{3}{4}\right) = 4\left(2 + \dfrac{3}{4}\right)$

$\qquad = 4 \cdot 2 + 4 \cdot \dfrac{3}{4}$

$\qquad = 8 + 3$

$\qquad = 11$

31. $6\left(3\dfrac{5}{6}\right) = 6\left(3 + \dfrac{5}{6}\right)$

$\qquad = 6 \cdot 3 + 6 \cdot \dfrac{5}{6}$

$\qquad = 18 + 5$

$\qquad = 23$

33. $-3(4)(-5) = -3(-5)(4)$
$\qquad -12(-5) = 15(4)$
$\qquad\qquad 60 = 60$
Commutative property of multiplication

35. $-3 + 4 + 5 = 4 + (-3) + 5$
$\qquad 1 + 5 = 1 + 5$
$\qquad\quad 6 = 6$
Commutative property of addition

37. $3 + (4 + 5) = (3 + 4) + 5$
$\qquad 3 + 9 = 7 + 5$
$\qquad 12 = 12$
Associative property of addition

39. a. $6(x + 2) = 6 \cdot x + 6 \cdot 2 = 6x + 12$

b. $-3(x - 3) = -3 \cdot x + (-3)(-3)$
$\qquad = -3x + 9$

c. $-6(x + 4) = -6 \cdot x + (-6) \cdot 4$
$\qquad = -6x - 24$

41. a. $-3(x + y - 5)$
$\qquad = -3 \cdot x + (-3) \cdot y + (-3)(-5)$
$\qquad = -3x - 3y + 15$

b. $-(x - y - z)$
$\qquad = -1 \cdot x + (-1)(-y) + (-1)(-z)$
$\qquad = -x + y + z$

43. a. $-(x - 3) = -1 \cdot x + (-1)(-3)$
$\qquad\qquad = -x + 3$

b. $y(4 + y) = y \cdot 4 + y \cdot y$
$\qquad\qquad = 4y + y^2$

c. $-(2 - y) = -1 \cdot 2 + (-1)(-y)$
$\qquad\qquad = -2 + y$

45. a. $\dfrac{-2x}{-6x} = \dfrac{-1 \cdot 2 \cdot x}{-1 \cdot 2 \cdot 3 \cdot x} = \dfrac{1}{3}$

b. $\dfrac{-14a}{21a} = \dfrac{-1 \cdot 2 \cdot 7 \cdot a}{3 \cdot 7 \cdot a} = \dfrac{-2}{3} = -\dfrac{2}{3}$

c. $\dfrac{6x}{15xyz} = \dfrac{2 \cdot 3 \cdot x}{3 \cdot 5 \cdot x \cdot y \cdot z} = \dfrac{2}{5yz}$

47. a. $\dfrac{15xy}{21y} = \dfrac{3 \cdot 5 \cdot x \cdot y}{3 \cdot 7 \cdot y} = \dfrac{5x}{7}$

b. $\dfrac{39abc}{13acd} = \dfrac{3 \cdot 13 \cdot a \cdot b \cdot c}{13 \cdot a \cdot c \cdot d} = \dfrac{3b}{d}$

c. $\dfrac{-12xy}{48xz} = \dfrac{-1 \cdot 2 \cdot 2 \cdot 3 \cdot x \cdot y}{2 \cdot 2 \cdot 2 \cdot 2 \cdot 3 \cdot x \cdot z} = \dfrac{-y}{4z}$

49. a. $\dfrac{ab}{bc} = \dfrac{a \cdot b}{b \cdot c} = \dfrac{a}{c}$

b. $\dfrac{ab}{ac} = \dfrac{a \cdot b}{a \cdot c} = \dfrac{b}{c}$

c. $\dfrac{ay}{by} = \dfrac{a \cdot y}{b \cdot y} = \dfrac{a}{b}$

51. a. $\dfrac{3x+4}{4} = \dfrac{3x}{4} + \dfrac{4}{4} = \dfrac{3}{4}x + 1$

b. $\dfrac{4x+8}{4} = \dfrac{4x}{4} + \dfrac{8}{4} = x + 2$

    or

$\dfrac{4x+8}{4} = \dfrac{4(x+2)}{4} = x + 2$

c. $\dfrac{x^2 + xy}{x} = \dfrac{x^2}{x} + \dfrac{xy}{x} = x + y$

    or

$\dfrac{x^2 + xy}{x} = \dfrac{x(x+y)}{x} = x + y$

d. $\dfrac{ab - bc}{b} = \dfrac{ab}{b} - \dfrac{bc}{b} = a - c$

    or

$\dfrac{ab - bc}{b} = \dfrac{b(a-c)}{b} = a - c$

53. $2a^2 + 4ab + 6b^2$

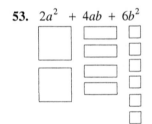

55. $3x^2 + 3x + 6$

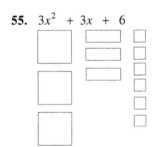

57. $9x$

59. a. -4

b. 1

c. -1

61. a. $4a^2 + 5a^2 = (4+5)a^2 = 9a^2$

b. No like terms

c. $4a^2 - 5a^2 = (4-5)a^2$
$$= -1a^2 \text{ or } -a^2$$

63. a. $-3x + 4x - 6x + 8x = 3x$

b. $-6y^2 + 8y^2 - 10y^2 - 3y^2 = -11y^2$

c. $2a + 6 + 3a + 9 = 2a + 3a + 6 + 9 = 5a + 15$

d. $3(x+1) - 2(x-1) =$
$3x + 3 - 2x + 2 =$
$3x - 2x + 3 + 2 = x + 5$

e. $3(x-4) + 4(4-x) =$
$3x - 12 + 16 - 4x =$
$3x - 4x - 12 + 16 = -x + 4$

f. $2x - 3y + 2x - 3y =$
$2x + 2x - 3y - 3y = 4x - 6y$

65. a. $\frac{1}{2}x + \frac{1}{4}y + \frac{1}{2}y - \frac{1}{4}x =$
$\frac{1}{2}x - \frac{1}{4}x + \frac{1}{4}y + \frac{1}{2}y =$
$\frac{1}{4}x + \frac{3}{4}y$

b. $0.5a + 0.75b - 0.5b + 1.5a =$
$0.5a + 1.5a + 0.75b - 0.5b$
$2a + 0.25b$

c. $-2x + 3y + (-4x) + (-6x) =$
$-2x - 4x - 6x + 3y = -12x + 3y$

d. $2(b+c) - 2(b-c) =$
$2b + 2c - 2b + 2c =$
$2b - 2b + 2c + 2c = 4c$

67. $x \cdot y = y \cdot x$

69. $a(b+c)$ is the product of two factors; applying the distributive property gives $ab + bc$, the sum of two terms.

71. a. $8 - (5 - 3) = 8 - 2 = 6$

b. $(8 - 5) - 3 = 3 - 3 = 0$

c. $16 \div (4 \div 2) = 16 \div 2 = 8$

d. $(16 \div 4) \div 2 = 4 \div 2 = 2$

e. No, subtraction is not associative.

f. No, division is not associative.

73. b and c need to be added or subtracted.

75. $3 + (4 \cdot 5) = (3 + 4) \cdot 5$

$\qquad 3 + 20 = 7 \cdot 5$

$\qquad\qquad 23 = 35 \;\text{ false}$

$23 \neq 35$; associative properties do not apply.

77. $3(4 \cdot 5) = 3 \cdot 4 \cdot 3 \cdot 5$

$\qquad 3 \cdot 20 = 12 \cdot 3 \cdot 5$

$\qquad\quad 60 = 36 \cdot 5$

$\qquad\quad 60 = 180 \;\text{ false}$

$60 \neq 180$; distributive property does not apply.

79. a. 3

$\quad$ **b.** 2

$\quad$ **c.** 4

$\quad$ **d.** 3

$\quad$ **e.** 3

81. a. $45 = 3 \cdot 3 \cdot 5$

$\quad$ **b.** 59 is prime.

$\quad$ **c.** $72 = 2 \cdot 2 \cdot 2 \cdot 3 \cdot 3$

$\quad$ **d.** $111 = 3 \cdot 37$

83. In like terms variables and exponents are identical.

85. Factors are multiplied; terms are added.

87. Two squares of area $x^2 = 2x^2$ (added). There is no multiplication of x^2 by x^2.

Mid-Chapter 2 Test

1. a. $2 - 5 = -3$

$\quad$ **b.** $-3 + 5 = 2$

$\quad$ **c.** $-3 - (-5) = -3 + 5 = 2$

2. a. $3 - (-4) = 3 + 4 = 7$

$\quad$ **b.** $-2 + (-5) = -7$

$\quad$ **c.** $6 + (-2.5) = 3.5$

3. a. $(-3)(2) = -6$

$\quad$ **b.** $(-5)(-3) = 15$

$\quad$ **c.** $-(-4) = (-1)(-4) = 4$

4. a. $\dfrac{27}{-3} = \dfrac{3 \cdot 9}{-3} = -9$

$\quad$ **b.** $\dfrac{-28}{-2} = \dfrac{-2 \cdot 14}{-2} = 14$

$\quad$ **c.** $\dfrac{-32}{4} = \dfrac{-4 \cdot 8}{4} = -8$

5. a. $4x + 5y - 2x + y =$
$\qquad 4x - 2x + 5y + y = 2x + 6y$

$\quad$ **b.** $2x^2 - 3x + 2x(1 - x) =$
$\qquad 2x^2 - 3x + 2x - 2x^2 =$
$\qquad 2x^2 - 2x^2 - 3x + 2x = -x$

6. a. $\dfrac{4xyz}{xy} = 4z$

$\quad$ **b.** $\dfrac{3x}{xyz} = \dfrac{3}{yz}$

$\quad$ **c.** $\dfrac{-2y}{4xy} = \dfrac{-2y}{2 \cdot 2xy} = -\dfrac{1}{2x}$

7. a. $-5.50 + 18.98 - 12.76 =$
$\qquad -5.50 - 12.76 + 18.98 =$
$\qquad -18.26 + 18.98 = 0.72$

$\quad$ **b.** $-3.89 - 42.39 + 50.00 =$
$\qquad -46.28 + 50.00 = 3.72$

8. $-3 + 6 + (-24) + 27$
$\quad = -3 - 24 + 27 + 6$
$\quad = -27 + 27 + 6 = 6$

9. $-6 + 12 + 18 - 24$
$\quad = -6 - 24 + 12 + 18$
$\quad = -30 + 30 = 0$

10. $13 \cdot (-4) \cdot 5 \cdot (-3)$
$= -3(13) \cdot (-4)(5)$
$= (-39)(-20) = 780$

11. a. $\dfrac{4x+2y}{2} = \dfrac{4x}{2} + \dfrac{2y}{2} = 2x + y$

$\dfrac{4x+2y}{2} = \dfrac{2(2x+y)}{2} = 2x + y$

b. $\dfrac{ab+bc}{b} = \dfrac{ab}{b} + \dfrac{bc}{b} = a + c$

$\dfrac{ab+bc}{b} = \dfrac{b(a+c)}{b} = a + c$

12.

x	$y = x - 1$
-2	-2 - 1 = -3
-1	-1 - 1 = -2
0	0 - 1 = -1
1	1 - 1 = 0
2	2 - 1 = 1
3	3 - 1 =2

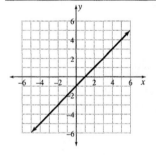

13.

x	$y = x + 1$
-2	-2 + 1 = -1
-1	-1 + 1 = 0
0	0 + 1 = 1
1	1 + 1 = 2
2	2 + 1 = 3
3	3 + 1 =4

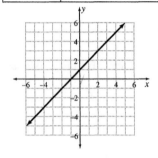

14.

x	$y = 2x - 3$
-2	2(-2) - 3 = -7
-1	2(-1) - 3 = -5
0	2(0) - 3 = -3
1	2(1) - 3 = -1
2	2(2) - 3 = 1
3	2(3) - 3 = 3

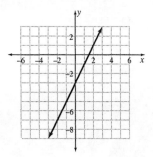

15.

x	$y = 3 - x$
-2	$3 - (-2) = 5$
-1	$3 - (-1) = 4$
0	$3 - 0 = 3$
1	$3 - 1 = 2$
2	$3 - 2 = 1$
3	$3 - 3 = 0$

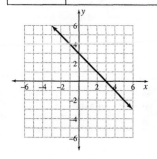

16. 8850 m - (-400 m) = 8850 m + 400 m
= 9250 m

17. 6194 m - (-86 m) = 6194 m + 86 m
= 6280 m

18. Mauna Kea's height from the ocean floor is;
13,710 ft - (-16,400 ft) = 13,710 ft + 16,400 ft
= 30,110 ft.

It is the "taller" mountain by:
30,110 ft - 29,028 ft = 1082 ft.

19. Subtraction is not commutative.

Exercises 2.4

1. **a.** In $3x^2$ the base is x.
$3x^2 = (3)(x)(x)$;
3 times the square of x

    b. In $-3x^2$ the base is x.
$-3x^2 = (-3)(x)(x)$;
The opposite of 3 times the square of x

    c. In $(-3x)^2$ the base is $(-3x)$.
$(-3x)^2 = (-3x)(-3x)$;
The square of the opposite of $3x$

    d. In ax^2 the base is x.
$ax^2 = (a)(x)(x)$;
a times x squared

    e. In $-x^2$ the base is x.
$-x^2 = (-1)(x)(x)$;
The opposite of x squared

    f. In $(-x^2)$ the base is $(-x)$.
$(-x)^2 = (-x)(-x)$;
The square of the opposite of x

3. **a.** $3^5 = 3 \cdot 3 \cdot 3 \cdot 3 \cdot 3 = 243$

    b. $2^6 = 2 \cdot 2 \cdot 2 \cdot 2 \cdot 2 \cdot 2 = 64$

    c. $(-2)^2 = (-2)(-2) = 4$

    d. $(-3)^3 = (-3)(-3)(-3) = -27$

5. **a.** $\left(\frac{1}{3}\right)^3 = \left(\frac{1}{3}\right)\left(\frac{1}{3}\right)\left(\frac{1}{3}\right) = \frac{1}{27}$

    b. $\left(\frac{4}{5}\right)^3 = \left(\frac{4}{5}\right)\left(\frac{4}{5}\right)\left(\frac{4}{5}\right) = \frac{64}{125}$

    c. $\left(-\frac{2}{3}\right)^3 = \left(-\frac{2}{3}\right)\left(-\frac{2}{3}\right)\left(-\frac{2}{3}\right) = -\frac{8}{27}$

    d. $-\left(-\frac{1}{3}\right)^2 = (-1)\left(-\frac{1}{3}\right)\left(-\frac{1}{3}\right) = -\frac{1}{9}$

7. **a.** $2 \cdot 4^2 = 2 \cdot 4 \cdot 4 = 32$

    b. $-2^2 = (-1) \cdot 2 \cdot 2 = -4$

    c. $3(-2)^2 = 3(-2)(-2) = 12$

    d. $-4 \cdot 3^2 = -4 \cdot 3 \cdot 3 = -36$

9. **a.** $x \cdot x \cdot x = x^3$

b. $x^3 x^2 = x \cdot x \cdot x \cdot x \cdot x = x^5$

c. $\left(a^3\right)^3 = a^3 \cdot a^3 \cdot a^3$
$$= a \cdot a \cdot a \cdot a \cdot a \cdot a \cdot a \cdot a \cdot a$$
$$= a^9$$

d. $b\left(b^2\right) = b\left(b \cdot b\right) = b \cdot b \cdot b = b^3$

11. a. $\dfrac{x^3}{x^4} = \dfrac{x \cdot x \cdot x}{x \cdot x \cdot x \cdot x} = \dfrac{x \cdot x \cdot x}{x \cdot x \cdot x} \cdot \dfrac{1}{x} = \dfrac{1}{x}$

b. $\dfrac{a^4}{a^3} = \dfrac{a \cdot a \cdot a \cdot a}{a \cdot a \cdot a} = \dfrac{a \cdot a \cdot a}{a \cdot a \cdot a} \cdot \dfrac{a}{1} = a$

c. $\dfrac{x^2 y}{\left(xy\right)^2} = \dfrac{x^2 y}{xy \cdot xy} = \dfrac{x \cdot x \cdot y}{x \cdot y \cdot x \cdot y}$
$$= \dfrac{x \cdot x \cdot y}{x \cdot x \cdot y} \cdot \dfrac{1}{y} = \dfrac{1}{y}$$

d. $\dfrac{a^3 b}{ab^2} = \dfrac{a \cdot a \cdot a \cdot b}{a \cdot b \cdot b} = \dfrac{a \cdot b}{a \cdot b} \cdot \dfrac{a \cdot a}{b} = \dfrac{a^2}{b}$

13. a. $\left(-4x\right)^2 = \left(-4\right)^2 x^2 = 16x^2$

b. $\left(3x\right)^3 = 3^3 x^3 = 27x^3$

c. $\left(0.2x\right)^2 = \left(0.2\right)^2 x^2 = 0.04x^2$

d. $-\left(2x\right)^3 = -2^3 x^3 = -8x^3$

15. a. $\left(ab\right)^2 = a^2 b^2$

b. $\left(xy\right)^3 = x^3 y^3$

c. $\left(2ac\right)^2 = 2^2 a^2 c^2 = 4a^2 c^2$

17. a. $-2\left(4a\right)^2 = -2 \cdot 4^2 a^2$
$$= -2 \cdot 16a^2$$
$$= -32a^2$$

b. $-3\left(-3x\right)^2 = \left(-3\right)\left(-3\right)^2 x^2$
$$= \left(-3\right)^3 x^2$$
$$= -27x^2$$

c. $-2\left(-2x\right)^3 = \left(-2\right)\left(-2\right)^3 x^3$
$$= \left(-2\right)^4 x^3$$
$$= 16x^3$$

19. a. $\left(\dfrac{x}{4}\right)^3 = \dfrac{x^3}{4^3} = \dfrac{x^3}{64}$

b. $\left(\dfrac{2n}{5}\right)^2 = \dfrac{\left(2n\right)^2}{5^2} = \dfrac{2^2 n^2}{25} = \dfrac{4n^2}{25}$

c. $\left(-\dfrac{3x}{5}\right)^2 = \dfrac{\left(3x\right)^2}{5^2} = \dfrac{3^2 x^2}{25} = \dfrac{9x^2}{25}$

21. a. $\left(\dfrac{4n}{3}\right)^2 = \dfrac{\left(4n\right)^2}{3^2} = \dfrac{4^2 n^2}{9} = \dfrac{16n^2}{9}$

b. $-\left(\dfrac{3x}{4}\right)^3 = -\dfrac{\left(3x\right)^3}{4^3} = -\dfrac{27x^3}{64}$

c. $\left(-\dfrac{5c}{3}\right)^3 = \dfrac{\left(-5c\right)^3}{3^3} = \dfrac{-125c^3}{27}$

23. a. $\left(8 \text{ in.}\right)^2 = 64 \text{ in.}^2$

b. $\left(0.3 \text{ sec}\right)^2 = 0.09 \text{ sec}^2$

c. $\left(6 \text{ m}\right)^3 = 216 \text{ m}^3$

25. a. $\dfrac{3 \text{ yd}}{6 \text{ yd}^3} = \dfrac{3 \text{ yd}}{6 \text{ yd} \cdot \text{yd} \cdot \text{yd}} = \dfrac{1}{2 \text{ yd}^2}$

b. $\dfrac{7 \text{ m}^3}{42 \text{ m}} = \dfrac{7 \text{ m} \cdot \text{m} \cdot \text{m}}{42 \text{ m}} = \dfrac{\text{m}^2}{6} = \dfrac{1}{6} \text{ m}^2$

c. $\dfrac{10 \text{ in.}^2}{\left(4 \text{ in.}\right)^3} = \dfrac{10 \text{ in.} \cdot \text{in.}}{\left(4 \text{ in.}\right)\left(4 \text{ in.}\right)\left(4 \text{ in.}\right)}$
$$= \dfrac{5}{32 \text{ in.}}$$

27. even exponents

29. a. $1 + 9 \cdot 9 - 8 = 1 + 81 - 8 = 74$

b. $1 + \sqrt{9} \cdot 9 - 8 = 1 + 3 \cdot 9 - 8 =$
$1 + 27 - 8 = 20$

  c. $\left(1+\sqrt{9}\right)\cdot 9 - 8 = (1+3)\cdot 9 - 8 =$

$\quad\quad 4\cdot 9 - 8 = 36 - 8 = 28$

31. $2\cdot 3\cdot 4 + 2\cdot 4\cdot 5 + 2\cdot 3\cdot 5 =$
$24 + 40 + 30 = 94$

33. $\frac{1}{2}\cdot 5\cdot(6+8) = \frac{1}{2}\cdot 5\cdot 14 =$
$\frac{1}{2}\cdot 14\cdot 5 = 7\cdot 5 = 35$

35. $6^2 + 8^2 = 36 + 64 = 100$

37. $5^2 + 12^2 = 25 + 144 = 169$

39. $7 - 2(4-1) = 7 - 2\cdot 3 = 7 - 6 = 1$

41. $(7-2)(4-1) = 5\cdot 3 = 15$

43. $6 - 3(x-4) = 6 - 3x + 12 =$
$-3x + 6 + 12 = -3x + 18$

45. $(6-3)(x-4) = 3(x-4) = 3x - 12$

47. $3[8 - 2(3-5)] = 3[8 - 2(-2)] =$
$3(8+4) = 3\cdot 12 = 36$

49. $|4-6| + |6-4| = |-2| + |2| = 2 + 2 = 4$

51. $|5-2| + |2-5| = |3| + |-3| = 3 + 3 = 6$

53. a. $\dfrac{4-7}{2-(-3)} = \dfrac{-3}{2+3} = \dfrac{-3}{5} = -\dfrac{3}{5}$

    b. $\dfrac{-2-3}{-2-(-3)} = \dfrac{-5}{-2+3} = \dfrac{-5}{1} = -5$

    c. $\dfrac{5-2}{3-(-4)} = \dfrac{3}{3+4} = \dfrac{3}{7}$

55. a. $\dfrac{3+\sqrt{25}}{4} = \dfrac{3+5}{4} = \dfrac{8}{4} = 2$

    b. $\dfrac{9-\sqrt{36}}{2} = \dfrac{9-6}{2} = \dfrac{3}{2}$

    c. $\dfrac{12+\sqrt{64}}{5} = \dfrac{12+8}{5} = \dfrac{20}{5} = 4$

57. $\sqrt{(-6-6)^2 + (12-3)^2} =$
$\sqrt{(-12)^2 + 9^2} = \sqrt{144 + 81} =$
$\sqrt{225} = 15$

59. $\sqrt{(4-(-2))^2 + (4-(-4))^2} =$
$\sqrt{(4+2)^2 + (4+4)^2} = \sqrt{6^2 + 8^2} =$
$\sqrt{36 + 64} = \sqrt{100} = 10$

61. $\dfrac{|3-7.5| + |9-7.5| + |10-7.5|}{3} =$

$\dfrac{|-4.5| + |1.5| + |2.5|}{3} =$

$\dfrac{4.5 + 1.5 + 2.5}{3} = \dfrac{8.5}{3} \approx 2.83$

63. $3.14(2)^2\cdot 6 = 3.14\cdot 4\cdot 6 = 75.36$

65. $\frac{4}{3}(3.14)\cdot 2^3 = \frac{4}{3}(3.14)\cdot 8 =$
$\dfrac{4\cdot 3.14\cdot 8}{3} = \dfrac{100.48}{3} \approx 33.49$

67. $2(3.14)\cdot 3\cdot 4 + 2(3.14)\cdot 3^2 =$
$6.28\cdot 12 + 6.28\cdot 9 \approx 131.88$

69. $-\frac{1}{2}(32)(-1)^2 + 16(-1) + 50 =$
$-16 + (-16) + 50 = 18$

71. One answer is to complete all arithmetic possible in an expression.

73. 4 was incorrectly subtracted from 7 first, before multiplication.

75. $\left(x^3\right)^2 = x^3\cdot x^3 = x^{3+3} = x^6$, not x^9

77. $\left(x^3\right)^2 = x^3\cdot x^3 = x^{3+3} = x^6$
and
$(-0.2)^2 = 0.04$

79. $\left(x^3\right)^2 = x^3\cdot x^3 = x^{3+3} = x^6$

81.

Bounce #	Height (cm)
0	96
1	$96 \div 2 = 48$
2	$48 \div 2 = 24$
3	$24 \div 2 = 12$
4	$12 \div 2 = 6$
5	$6 \div 2 = 3$

The ball will bounce 3 cm on the fifth bounce.

Exercises 2.5

1. $\dfrac{8 \text{ oz} \cdot 1 \text{ lb}}{16 \text{ oz}} = \dfrac{1}{2} \text{ lb}$

3. $\dfrac{100 \text{ yd} \cdot 3 \text{ ft} \cdot 12 \text{ in}}{1 \text{ yd} \cdot 1 \text{ ft}} = 100 \cdot 3 \cdot 12 \text{ in}$

$\qquad = 3600 \text{ in}$

5. $\dfrac{\text{feet}^3}{\text{foot}} = \dfrac{\text{foot} \cdot \text{feet}^2}{\text{foot}} = \text{feet}^2$

7. $\dfrac{\text{inches}^2}{\text{inches}^3} = \dfrac{\text{inches}^2}{\text{inches}^2 \cdot \text{inch}} = \dfrac{1}{\text{inch}}$

9. $\dfrac{1 \text{ km}}{1} \cdot \dfrac{1000 \text{ m}}{1 \text{ km}} \cdot \dfrac{39.37 \text{ in}}{1 \text{ m}} \cdot \dfrac{1 \text{ft}}{12 \text{ in}}$

$\approx 3281 \text{ ft}$

11. $\dfrac{120 \text{ lb}}{1} \cdot \dfrac{1 \text{ kg}}{2.2 \text{ lb}} \cdot \dfrac{1000 \text{ g}}{1 \text{ kg}} \approx 54{,}545 \text{ g}$

13. $\dfrac{1 \text{ day}}{1} \cdot \dfrac{24 \text{ hrs}}{1 \text{ day}} \cdot \dfrac{60 \text{ min}}{1 \text{ hr}} \cdot \dfrac{60 \text{ sec}}{1 \text{ min}}$

$= 86{,}400 \text{ sec}$

15. $\dfrac{1{,}000{,}000 \text{ sec}}{1} \cdot \dfrac{1 \text{ min}}{60 \text{ sec}} \cdot \dfrac{1 \text{ hr}}{60 \text{ min}} \cdot \dfrac{1 \text{ day}}{24 \text{ hr}}$

$\approx 11.6 \text{ days}$

17. $\dfrac{72 \text{ yr}}{1} \cdot \dfrac{365 \text{ days}}{1 \text{ yr}} \cdot \dfrac{86{,}400 \text{ sec}}{1 \text{ day}}$

$= 2{,}270{,}592{,}000 \text{ sec}$

19. $\dfrac{1 \text{ ft}^3}{1} \cdot \dfrac{12 \text{ in}}{1 \text{ ft}} \cdot \dfrac{12 \text{ in}}{1 \text{ ft}} \cdot \dfrac{12 \text{ in}}{1 \text{ ft}} = 1728 \text{ in}^3$

21. $\dfrac{1 \text{ yd}^2}{1} \cdot \dfrac{36 \text{ in}}{1 \text{ yd}} \cdot \dfrac{36 \text{ in}}{1 \text{ yd}} = 1296 \text{ in}^2$

23. $\dfrac{1200 \text{ in}^2}{1} \cdot \dfrac{1 \text{ ft}}{12 \text{ in}} \cdot \dfrac{1 \text{ ft}}{12 \text{ in}} = \dfrac{1200 \text{ ft}^2}{144}$

$\approx 8.3 \text{ ft}^2$

25. $\dfrac{19.77 \text{ km}}{1} \cdot \dfrac{3281 \text{ ft}}{1 \text{ km}} \approx 65{,}000 \text{ ft}$

27. a. $P = 1.8 \text{ cm} + 2.0 \text{ cm} + 2.1 \text{ cm} + 4.0 \text{ cm} = 9.9 \text{ cm}$

$A = \frac{1}{2}(1.6 \text{ cm})(2.0 \text{ cm} + 4.0 \text{ cm})$
$= (0.8)(6.0) \text{ cm}^2 = 4.8 \text{ cm}^2$

b. $P = 2.0 \text{ cm} + 3.0 \text{ cm} + 2.1 \text{ cm} + 5.5 \text{ cm} = 12.6 \text{ cm}$

$A = \frac{1}{2}(1.6 \text{ cm})(3.0 \text{ cm} + 5.5 \text{ cm})$
$= (0.8)(8.5) \text{ cm}^2 = 6.8 \text{ cm}^2$

c. Perimeter is circumference;
$C = 2\pi(2 \text{ ft}) = 4\pi \text{ ft} \approx 12.6 \text{ ft}$

$A = \pi(2 \text{ ft})^2 = 4\pi \text{ ft}^2 \approx 12.6 \text{ ft}^2$

d. $P = 3.5 \text{ cm} + 2.3 \text{ cm} + 3.8 \text{ cm}$
$= 9.6 \text{ cm}$

$A = \frac{1}{2}(3.8 \text{ cm})(2.1 \text{ cm})$
$= (1.9)(2.1) \text{ cm}^2 \approx 4.0 \text{ cm}^2$

29. a. $P = 26 \text{ yd} + 10 \text{ yd} + 24 \text{ yd}$
$= 60 \text{ yd}$

$A = \frac{1}{2}(24 \text{ yd})(10 \text{ yd})$
$= (12)(10) \text{ yd}^2 = 120 \text{ yd}^2$

b. $P = (4)(1.7 \text{ cm}) = 6.8 \text{ cm}$
$A = (1.7 \text{ cm})^2 \approx 2.9 \text{ cm}^2$

c. $P = 31.9 \text{ m} + 17 \text{ m} + 36.1 \text{ m}$
$= 85 \text{ m}$
$A = \frac{1}{2}(36.1 \text{ m})(15 \text{ m}) \approx 270.8 \text{ m}^2$

d. Perimeter is circumference
$C = \pi(9 \text{ in}) = 9\pi \text{ in} \approx 28.3 \text{ in}$
$A = \pi\left(\frac{9}{2} \text{ in}\right)^2 = \frac{81}{4}\pi \text{ in}^2 \approx 63.6 \text{ in}^2$

31. a. $S = 2(10 \text{ ft})(5 \text{ ft}) + 2(6 \text{ ft})(10 \text{ ft})$
$+ 2(6 \text{ ft})(5 \text{ ft})$
$= (100 + 120 + 60) \text{ ft}^2$
$= 280 \text{ ft}^2$

$V = (10 \text{ ft})(6 \text{ ft})(5 \text{ ft}) = 300 \text{ ft}^3$

b. $S = 2(12\text{ in})(5\text{ in}) + 2(4\text{ in})(12\text{ in})$
$\qquad + 2(5\text{ in})(4\text{ in})$
$\qquad = (120 + 96 + 40)\text{ in}^2 = 256\text{ in}^2$

$\qquad V = (12\text{ in})(5\text{ in})(4\text{ in}) = 240\text{ in}^3$

33. a. $S = 2\pi(6\text{ cm})^2 + 2\pi(6\text{ cm})(8\text{ cm})$
$\qquad = (72\pi + 96\pi)\text{ cm}^2 = 168\pi\text{ cm}^2$
$\qquad \approx 527.5\text{ cm}^2$

$\qquad V = \pi(6\text{ cm})^2(8\text{ cm}) = 288\pi\text{ cm}^3$
$\qquad \approx 904.3\text{ cm}^3$

b. $S = 2\pi(3\text{ cm})^2 + 2\pi(3\text{ cm})(4\text{ cm})$
$\qquad = (18\pi + 24\pi)\text{ cm}^2 = 42\pi\text{ cm}^2$
$\qquad \approx 131.9\text{ cm}^2$

$\qquad V = \pi(3\text{ cm})^2(4\text{ cm}) = 36\pi\text{ cm}^3$
$\qquad \approx 113.0\text{ cm}^3$

35. a. $S = 4\pi(8\text{ cm})^2 = 256\pi\text{ cm}^2$
$\qquad \approx 803.8\text{ cm}^2$

$\qquad V = \frac{4}{3}\pi(8\text{ cm})^3 = \frac{2048}{3}\pi\text{ cm}^3$
$\qquad \approx 2143.6\text{ cm}^3$

b. $S = 4\pi(4\text{ cm})^2 = 64\pi\text{ cm}^2$
$\qquad \approx 201.0\text{ cm}^2$
$\qquad V = \frac{4}{3}\pi(4\text{ cm})^3 = \frac{256}{3}\pi\text{ cm}^3$
$\qquad \approx 267.9\text{ cm}^3$

37. The area of a 9" pizza is:
$\pi(9\text{ in})^2 = 81\pi\text{ in}^2$
The area of a 6" pizza is:
$\pi(6\text{ in})^2 = 36\pi\text{ in}^2$
$\dfrac{81\pi\text{ in}^2}{36\pi\text{ in}^2} = \dfrac{81}{36} = 2.25$

The large pizza has an area that is 2.25 times bigger than the area of the small pizza.

39. a. $P = 2x \cdot 4 = 8x;\quad A = (2x)^2 = 4x^2$

b. $P = 4x \cdot 4 = 16x;\quad A = (4x)^2 = 16x^2$

c. $\dfrac{8x}{16x} = \dfrac{1}{2}$

d. $\dfrac{4x^2}{16x^2} = \dfrac{1}{4}$

41. $200\text{ serv.} \cdot \dfrac{12\text{ oz}}{1\text{ serv}} \cdot \dfrac{1\text{ cup}}{8\text{ oz}} \cdot \dfrac{1\text{ qt}}{4\text{ cup}} \cdot \dfrac{1\text{ gal}}{4\text{ qt}}$

$\qquad \approx 19\text{ gal}$

43. $V = (4.5\text{ in})(7.0\text{ in})(13.5\text{ in})$
$\qquad \approx 425.25\text{ in}^3$

$\qquad (1000)(425.25)\text{ in}^3 = 425{,}250\text{ in}^3$

$\qquad 425{,}250\text{ in}^3 \cdot \left(\dfrac{1\text{ ft}}{12\text{ in}}\right)^3$

$\qquad = \dfrac{425{,}250\text{ in}^3 \cdot 1\text{ ft}^3}{1728\text{ in}^3} \approx 246.1\text{ ft}^3$

45. $\dfrac{12\text{ oz}}{1\text{ spill}} \cdot \dfrac{1\text{ cup}}{8\text{ oz}} \cdot \dfrac{1\text{ qt}}{4\text{ cups}} \cdot \dfrac{1\text{ gal}}{4\text{ qt}} \cdot$

$\qquad \dfrac{231\text{ in}^3}{1\text{ gal}} \cdot \dfrac{1\text{ spill}}{\frac{1}{16}\text{ in}} = 346.5\text{ in}^2$

47. a. $2{,}500{,}000\text{ gal} \cdot \dfrac{231\text{ in}^3}{1\text{ gal}} \cdot \dfrac{1\text{ ft}^3}{12^3\text{ in}^3} \cdot$

$\qquad \dfrac{1\text{ width}}{40\text{ ft}} \cdot \dfrac{1\text{ thickness}}{6\text{ ft}} \approx 1392.5\text{ ft}$

b. $V = \pi r^2 h = \pi\left(\dfrac{100}{2}\text{ ft}\right)^2(50\text{ ft})$

$\qquad = 125{,}000\pi\text{ ft}^3$
$\qquad \approx 392{,}500\text{ ft}^3$

$\qquad 392{,}500\text{ ft}^3 \cdot \dfrac{(12\text{ in.})^3}{1\text{ ft}^3} \cdot \dfrac{1\text{ gal}}{231\text{ in.}^3}$
$\qquad \approx 2{,}940{,}000\text{ gal}$

c. $2{,}500{,}000\text{ gal} \cdot \dfrac{231\text{ in}^3}{1\text{ gal}} \cdot \dfrac{1\text{ ft}^3}{12^3\text{ in}^3} \cdot$

$\qquad \dfrac{1\text{ depth}}{1\text{ in}} \approx 4{,}010{,}417\text{ ft}^2$

$\qquad \approx 0.14\text{ mi}^2$

d. $4{,}010{,}417\text{ ft}^2 \cdot \dfrac{1\text{ mi}^2}{(5280\text{ ft})^2} \cdot \dfrac{640\text{ acres}}{1\text{ mi}^2}$

$\qquad \approx 92.1\text{ acres}$

49. Add the lengths of the parallel sides, multiply by the height, and then divide by 2.

51. Multiply the length, width, and height together.

53. We add 2 lengths and 2 widths.

55. Blue area is a 3 cm square minus a 3 cm diameter circle.

$A = (3 \text{ cm})^2 - \pi(\frac{3}{2} \text{ cm})^2 = (9 - \frac{9}{4}\pi) \text{ cm}^2 \approx 1.9$ cm^2

57. Inside area is a 100 m by 73 m rectangle plus 2 half circles of 73 m diameter.

$A = (100 \text{ m})(73 \text{ m}) + \pi(\frac{73}{2} \text{ m})^2$
$= (7300 + \frac{5329}{4}\pi) \text{ m}^2 \approx 11{,}483.3 \text{ m}^2$

59. Shaded area is a 15 m by 9 m rectangle minus a (15-12) m by (9-3) m rectangle.

$A = (15 \text{ m})(9 \text{ m}) - (3 \text{ m})(6 \text{ m})$
$= (135 - 18) \text{ m}^2 = 117 \text{ m}^2$

Exercises 2.6

1.

3. a. $-8 < -3$

b. $+4 > -9$

c. $(-3)^2 = 3^2$

d. $0.5 > 0.5^2$

e. $6 > -5$

f. $(-2)(6) < (-2)(-5)$
$-12 < 10$

g. $-6 < -5$

h. $(-2)(-6) > (-2)(-5)$
$12 > 10$

5. a. $-3.75 < -3.25$

b. $3(-2) = -3(2)$

c. $\frac{1}{2} > -\frac{1}{2}$

d. $|-4| > |2|$
$4 > 2$

e. $(-2)(-3) > 2(-4)$
$6 > -8$

f. $\left(\frac{1}{2}\right)^2 = \left(-\frac{1}{2}\right)^2$

g. $-2.5 > -3$

h. $\frac{22}{7} > \pi$

7. a. $0 < x < 4$

b. $-5 < x \le -2$

9. a. $x > 3$ and $x < 8$

b. $x > -3$ and $x \le -1$

c. $x > -2$ and $x < 1$

11. f, s; $x < 3$, $(-\infty, 3)$

13. e, w; $x \ge 3$, $[3, +\infty)$

15. b, r; $-3 \le x \le 3$, $[-3, 3]$

17. a. $[-1, 3)$;
x is between -1 and 3, including -1;

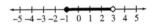

b. $-4 < x \le -1$; x is greater than -4 and less than or equal to -1;

c. $-3 \le x < 5$; $[-3, 5)$;

d. $(-\infty, -4)$; x is less than -4;

e. $-2 \le x < 5$; $[-2, 5)$;
x is between -2 and 5, including –2

f. $(-2, +\infty)$; x is greater than -2;

g. $-4 < x \le 2$; $(-4, 2]$;

h. $x \ge -3$; x is greater than or equal to -3;

19. A statement that one quantity is greater than or less than another quantity.

21. An interval is a set of numbers between endpoints that may also include one or both endpoints.

23. x cannot be less than 2 <u>and</u> greater than 4.

25. 2 is equal to 2 and $\le$ includes the possibility of equality.

27. **a.** $x \le 2000$

  b. $2000 < x \le 5000$

  c. $x > 5000$

29. **a.** $x \le -5$

  b. $-5 < x < 5$

  c. $x \ge 5$

31. **a.** $x < 5$

  b. $5 \le x \le 50$

  c. $x > 50$

33.

$n\ \%$	\$1.00	\$5.00	\$10.00	x
6	$0.06(1) = \$0.06$	$0.06(5) = \$0.30$	$0.06(10) = \$0.60$	$0.06x$
10	$0.10(1) = \$0.10$	$0.10(5) = \$0.50$	$0.10(10) = \$1.00$	$0.10x$
25	$0.25(1) = \$0.25$	$0.25(5) = \$1.25$	$0.25(10) = \$2.50$	$0.25x$
100	$1(1) = \$1.00$	$1(5) = \$5.00$	$1(10) = \$10.00$	x
150	$1.5(1) = \$1.50$	$1.5(5) = \$7.50$	$1.5(10) = \$15.00$	$1.5x$

Chapter 2 Review

1. **a.** $-3 + (-5) = -8$

  b. $3 + (-8) = -5$

  c. $4 - 17 = 4 + (-17) = -13$

  d. $-5 - (-18) = -5 + (+18) = 13$

  e. $-21 + 7 = -14$

  f. $-26 + 19 = -7$

  g. $14 - (-28) = 14 + (+28) = 42$

  h. $12 - 36 = 12 + (-36) = -24$

  i. $-32 - (-16) = -32 + (+16) = -16$

  j. $-4 - (-16) = -4 + (+16) = 12$

  k. $-11 + 22 = 11$

  l. $8 - (-5) = 8 + (+5) = 13$

3. **a.** $(-9)(6) = -54$

  b. $(-9)(-6) = 54$

  c. $(-18)(-3) = 54$

  d. $(-18)(3) = -54$

  e. $(-8)(-7) = 56$

  f. $(8)(-7) = -56$

  g. $4(-14) = -56$

  h. $(-4)(-14) = 56$

  i. $(-48) \div (-24) = 2$

  j. $(-48) \div 12 = -4$

  k. $48 \div (-6) = -8$

  l. $(-48) \div (-6) = 8$

  m. $(-48) \div (-3) = 16$

  n. $(-48) \div 3 = -16$

  o. $48 \div (-8) = -6$

5. a. $8+7+2+3=(8+2)+(7+3)$
$$=10+10$$
$$=20$$

b. $92+55+8+45$
$$=(92+8)+(55+45)$$
$$=100+100$$
$$=200$$

c. $1.25+2.69+3.75$
$$=(1.25+3.75)+2.69$$
$$=5+2.69$$
$$=7.69$$

d. $\$5.29+\$4.98+\$5.02$
$$=\$5.29+(\$4.98+\$5.02)$$
$$=\$5.29+\$10.00$$
$$=\$15.29$$

e. $2\frac{1}{2}+1\frac{1}{3}+4\frac{1}{4}+1\frac{1}{4}$
$$=(2+1+4+1)+\left(\frac{1}{2}+\frac{1}{4}+\frac{1}{4}+\frac{1}{3}\right)$$
$$=8+1+\frac{1}{3}$$
$$=9\frac{1}{3}$$

f. $3\frac{7}{8}+1\frac{1}{2}+2\frac{1}{8}+3\frac{1}{2}$
$$=(3+1+2+3)+\left(\frac{7}{8}+\frac{1}{8}+\frac{1}{2}+\frac{1}{2}\right)$$
$$=9+2$$
$$=11$$

7. a. $3\cdot5\cdot4=3\cdot(5\cdot4)=3\cdot(20)=60$

b. $8\cdot2\cdot5=8\cdot(2\cdot5)=8\cdot10=80$

c. $9\cdot4\cdot25=9\cdot(4\cdot25)=9\cdot100=900$

d. $\frac{1}{2}\cdot17\cdot20=17\cdot\left(\frac{1}{2}\cdot20\right)$
$$=17\cdot10$$
$$=170$$

e. $9\cdot\frac{1}{2}\cdot6=9\cdot\left(\frac{1}{2}\cdot6\right)=9\cdot3=27$

f. $5\cdot\frac{1}{4}\cdot8=5\cdot\left(\frac{1}{4}\cdot8\right)=5\cdot2=10$

9. a. $4\left(3\frac{1}{4}\right)=4\left(3+\frac{1}{4}\right)$
$$=4\cdot3+4\cdot\frac{1}{4}$$
$$=12+1$$
$$=13$$

b. $6\left(4\frac{1}{3}\right)=6\left(4+\frac{1}{3}\right)$
$$=6\cdot4+6\cdot\frac{1}{3}$$
$$=24+2$$
$$=26$$

c. $8\left(1\frac{3}{4}\right)=8\left(1+\frac{3}{4}\right)$
$$=8\cdot1+8\cdot\frac{3}{4}$$
$$=8+\frac{24}{4}$$
$$=8+6$$
$$=14$$

d. $3\left(2\frac{2}{3}\right)=3\left(2+\frac{2}{3}\right)$
$$=3\cdot2+3\cdot\frac{2}{3}$$
$$=6+\frac{6}{3}$$
$$=6+2$$
$$=8$$

e. $9\left(1\frac{2}{3}\right)=9\left(1+\frac{2}{3}\right)$
$$=9\cdot1+9\cdot\frac{2}{3}$$
$$=9+\frac{18}{3}$$
$$=9+6$$
$$=15$$

f.
$$8\left(2\frac{3}{8}\right) = 8\left(2 + \frac{3}{8}\right)$$
$$= 8 \cdot 2 + 8 \cdot \frac{3}{8}$$
$$= 16 + \frac{24}{8}$$
$$= 16 + 3$$
$$= 19$$

11. a. $2(x + y) = 2x + 2y$

b. $ac + ab = a(c + b)$

c. $4(x^2 - 2 + 3) = 4x^2 - 8x + 12$

d. $3xy + 4x^2 y = xy(3 + 4x)$

e. $3(2x + 4y - 5) = 6x + 12y - 15$

f. $15a^2 bc + 5ab^2 + 10abc =$
$\qquad 5ab(3ac + b + 2c)$

13. a. $\dfrac{abc}{bcd} = \dfrac{a}{d} \cdot \dfrac{bc}{bc} = \dfrac{a}{d}$

b. $\dfrac{4xy}{6xz} = \dfrac{2x \cdot 2y}{2x \cdot 3z} = \dfrac{2y}{3z}$

c. $\dfrac{-21cd}{14ad} = \dfrac{-7d \cdot 3c}{7d \cdot 2a} = -\dfrac{3c}{2a}$

d. $(-2x)^2 = (-2)^2 x^2 = 4x^2$

e. $(-3y)^3 = (-3)^3 y^3 = -27y^3$

f. $(-2y)^4 = (-2)^4 y^4 = 16y^4$

g. $(-ab)^2 = (-a)^2 b^2 = a^2 b^2$

h. $(ab)^2 = a^2 b^2$

i. $m^4 m^5 = m^{4+5} = m^9$

j. $m^2 m^7 = m^{2+7} = m^9$

k. $m^5 \div m^2 = m^{5-2} = m^3$

l. $m^7 \div m^4 = m^{7-4} = m^3$

m. $\dfrac{3x + 6y}{3} = \dfrac{3x}{3} + \dfrac{6y}{3} = x + 2y$

n. $\dfrac{mn + n^2}{n} = \dfrac{mn}{n} + \dfrac{n^2}{n} = m + n$

o. $\dfrac{2a + 4b}{4} = \dfrac{2a}{4} + \dfrac{4b}{b} = \tfrac{1}{2}a + b$

15. a. $\left(\dfrac{4x}{y}\right)^2 = \dfrac{4^2 x^2}{y^2} = \dfrac{16x^2}{y^2}$

b. $\left(\dfrac{x}{3y}\right)^3 = \dfrac{x^3}{3^3 y^3} = \dfrac{x^3}{27y^3}$

c. $\left(-3ab^2\right)^2 = (-3)^2 a^2 \left(b^2\right)^2$
$\qquad = 9a^2 b^{2+2}$
$\qquad = 9a^2 b^4$

d. $\dfrac{3x^3}{9x} = \dfrac{3 \cdot x \cdot x^2}{3 \cdot 3 \cdot x} = \dfrac{x^2}{3}$

e. $\dfrac{4ab^2}{a^3 b} = \dfrac{4 \cdot a \cdot b \cdot b}{a^2 \cdot a \cdot b} = \dfrac{4b}{a^2}$

f. $\dfrac{xy^3 z}{x^2 yz^2} = \dfrac{x \cdot y \cdot y^2 \cdot z}{x \cdot x \cdot y \cdot z \cdot z} = \dfrac{y^2}{xz}$

g. $\dfrac{(-x)^3}{-x^2} = \dfrac{(-1)^3 x^3}{-1 \cdot x \cdot x} = \dfrac{-1 \cdot x \cdot x \cdot x}{-1 \cdot x \cdot x} = x$

h. $\dfrac{-3x^3}{(9x)^2} = \dfrac{-3 \cdot x \cdot x \cdot x}{9 \cdot x \cdot 9 \cdot x}$
$\qquad = \dfrac{-3 \cdot x \cdot x \cdot x}{3 \cdot 3 \cdot x \cdot 3 \cdot 3 \cdot x}$
$\qquad = \dfrac{-x}{27}$
$\qquad = -\dfrac{1}{27}x$

i. $\dfrac{-4x^3}{(-4x)^2} = \dfrac{-4 \cdot x \cdot x \cdot x}{-4 \cdot x \cdot -4 \cdot x} = \dfrac{x}{-4} = -\dfrac{1}{4}x$

j. $\dfrac{2 \text{ qt}}{2 \text{ qt}} = 1$

k. $\dfrac{3 \text{ ft}}{1 \text{ yd}} = \dfrac{3 \text{ ft}}{1 \text{ yd}} \cdot \dfrac{1 \text{ yd}}{3 \text{ ft}} = \dfrac{3}{3} = 1$

l. $\dfrac{125 \text{ km}^3}{\left(5 \text{ km}\right)^2} = \dfrac{125 \text{ km} \cdot \text{km} \cdot \text{km}}{25 \text{ km} \cdot \text{km}} = 5 \text{ km}$

17. a. $-(-2)^2 = -4$

    b. $4 - (-2) + (-2)^2 = 4 + 2 + 4 = 10$

    c. $5 - (-3) + (-2)^2 = 5 + 3 + 4 = 12$

    d. $-(-3)^2 = -9$

    e. $\sqrt{(3^2 + 4^2)} = \sqrt{9 + 16} = \sqrt{25} = 5$

    f. $\sqrt{(8^2 + 6^2)} = \sqrt{64 + 36} = \sqrt{100} = 10$

    g. $\sqrt{(25^2 - 20^2)} = \sqrt{625 - 400}$
    $= \sqrt{225}$
    $= 15$

    h. $\sqrt{(15^2 - 12^2)} = \sqrt{225 - 144}$
    $= \sqrt{81}$
    $= 9$

    i. $\sqrt{(1.5^2 + 2^2)} = \sqrt{2.25 + 4} = \sqrt{6.25} = 2.5$

    j. $\sqrt{(10^2 - 6^2)} = \sqrt{100 - 36} = \sqrt{64} = 8$

    k. $-|-4| = -4$

    l. $|-6 - (-5)| = |-6 + 5| = |-1| = 1$

19. a. $A = \pi(2.5 \text{ ft})^2 = \pi(6.25) \text{ ft}^2$
    $\approx 19.6 \text{ ft}^2$

    b. $A = \frac{1}{2}(5 \text{ yd})(4 \text{ yd}) = 10 \text{ yd}^2$

    c. $V = \frac{4}{3}\pi(3 \text{ m})^3 = 4\pi(9) \text{ m}^3$
    $= 113.1 \text{ m}^3$

    d. $V = (1.5 \text{ cm})^3 \approx 3.4 \text{ cm}^3$

21. $\dfrac{16 \text{ oz}}{1 \text{ box}} \cdot \dfrac{1 \text{ box}}{32 \text{ serv}} \cdot \dfrac{1 \text{ serv}}{12 \text{ crackers}} \cdot \dfrac{140 \text{ cal}}{1 \text{oz}}$
$\approx \dfrac{5.8 \text{ cal}}{1 \text{ cracker}}$

23. a. $\dfrac{3+4}{5-8}$ or $(3+4) \div (5-8)$; h and j

    b. $3 + \dfrac{4}{5} - 8$ or $3 + 4 \div 5 - 8$; e and k

    c. $\dfrac{3+4}{5} - 8$ or $(3+4) \div 5 - 8$; f and l

    d. $3 + \dfrac{4}{5-8}$ or $3 + 4 \div (5-8)$; g and i

    e. Three plus the quotient of 4 and 5 is decreased by 8.

    f. The sum of 3 and 4 is divided by 5 and decreased by 8.

    g. Three is added to the quotient of 4 and the difference between 5 and 8.

    h. The sum of 3 and 4 is divided by the difference between 5 and 8.

    i. Three is added to the quotient of 4 and the difference between 5 and 8.

    j. The sum of 3 and 4 is divided by the difference between 5 and 8.

    k. Three plus the quotient of 4 and 5 is decreased by 8.

    l. The sum of 3 and is divided by 5 and decreased by 8.

25. a. $V = \pi r^2 h$
    $= \pi (6 \text{ ft})^2 (7 \text{ ft})$
    $= 252\pi \text{ ft}^3$
    $\approx 791 \text{ ft}^3$

    $S = 2\pi r^2 + 2\pi r h$
    $= 2\pi (6 \text{ ft})^2 + 2\pi (6 \text{ ft})(7 \text{ ft})$
    $= 72\pi \text{ ft}^2 + 84\pi \text{ ft}^2$
    $= 156\pi \text{ ft}^2$
    $\approx 490 \text{ ft}^2$

b. $V = \dfrac{4}{3}\pi r^3$

$\quad = \dfrac{4}{3}\pi (8 \text{ in.})^3$

$\quad = \dfrac{2048\pi}{3} \text{ in.}^3$

$\quad \approx 2144 \text{ in.}^3$

$S = 4\pi r^2$

$\quad = 4\pi (8 \text{ in.})^2$

$\quad = 256\pi \text{ in.}^2$

$\quad \approx 804 \text{ in.}^2$

c. $V = l \cdot w \cdot h$

$\quad = (8 \text{ in.})(2 \text{ in.})(5 \text{ in.})$

$\quad = 80 \text{ in.}^3$

$S = 2lw + 2lh + 2wh$

$\quad = 2(8 \text{ in.})(2 \text{ in.}) + 2(8 \text{ in.})(5 \text{ in.})$

$\quad\quad + 2(2 \text{ in.})(5 \text{ in.})$

$\quad = 32 \text{ in.}^2 + 80 \text{ in.}^2 + 20 \text{ in.}^2$

$\quad = 132 \text{ in.}^2$

27. a. $4 > -3$

b. $2(-3) < (-2)(-3),\ -6 < 6$

c. $(-2)^2 > -2^2,\ 4 > -4$

d. $|-4| = |4|,\ 4 = 4$

e. $-2^3 = (-2)^3,\ -8 = -8$

f. $|-5| > -|5|,\ 5 > -5$

g. $-\dfrac{1}{4} > -\dfrac{1}{2}$

h. $-1.3 > -1.5$

29. One possible example:
$(3 + 4) + 5 = 7 + 5 = 12$:
$3 + (4 + 5) = 3 + 9 = 12$

31. The small circle excludes the point and is used with < and >. The dot includes the point and is used with ≤ and ≥.

33. $-x^2$ is the opposite of $(x)(x)$ while $(-x)^2$ is $(-x)(-x)$.

35. Change $2\frac{1}{4}$ to $\frac{9}{4}$ and write $\frac{4}{9}$.

37. Change subtraction to addition of the opposite; change division to multiplication of the reciprocal.

39. Answers may vary.

41.

Input: Cost of item	Inequality	Interval
Less than $50	$0 \le x < \$50$	[0, 50)
$50 to $500	$50 \le x \le 500$	[50, 500]
Over $500	$x > 500$	(500, +∞)

Chapter 2 Test

1.

x	$y = 3 - x$
-2	$3 - (-2) = 5$
0	$3 - 0 = 3$
2	$3 - 2 = 1$
4	$3 - 4 = -1$

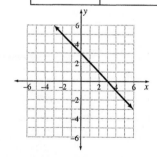

2. a. $\dfrac{3}{4} + \dfrac{5}{6} = \dfrac{9}{12} + \dfrac{10}{12} = \dfrac{19}{12}$

b. $\dfrac{3}{4} - \dfrac{5}{6} = \dfrac{9}{12} - \dfrac{10}{12} = -\dfrac{1}{12}$

c. $\dfrac{3}{4} \cdot \dfrac{-5}{6} = \dfrac{3(-5)}{4(6)} = \dfrac{-15}{24} = \dfrac{-5}{8}$

d. $\dfrac{-3}{4} \div \dfrac{5}{6} = \dfrac{-3}{4} \cdot \dfrac{6}{5} = \dfrac{(-3)(6)}{(4)(5)} = \dfrac{-18}{20} = \dfrac{-9}{10}$

3. Commutative property of addition.

4. Associative property of addition.

5. a. $-5 + 9 = 4$

b. $-1.4 + 2.5 - 3.6 = 2.5 - 5.0 = -2.5$

c. $-4 - (-3) = -4 + 3 = -1$

d. $(-3)(4)(-5) = -12(-5) = 60$

e. $8 - (-3)^2 = 8 - 9 = -1$

f. $\sqrt{26^2 - 24^2} = \sqrt{676 - 576} = \sqrt{100} = 10$

g. $m^2 m^5 = m \cdot m \cdot m \cdot m \cdot m \cdot m \cdot m = m^7$

h. $m^5 \div m^3 = \dfrac{m \cdot m \cdot m \cdot m \cdot m}{m \cdot m \cdot m} = m \cdot m = m^2$

i. $36 \div 2 \cdot 2 - 3 + (3^2 - 5) =$
$36 \div 2 \cdot 2 - 3 + (9 - 5) =$
$36 \div 2 \cdot 2 - 3 + 4 = 18 \cdot 2 - 3 + 4 =$
$36 - 3 + 4 = 37$

j. $\dfrac{ace}{aft} = \dfrac{a}{a} \cdot \dfrac{ce}{ft} = \dfrac{ce}{ft}$

k. $\dfrac{6x^2}{9x} = \dfrac{3 \cdot 2}{3 \cdot 3} x^{2-1} = \tfrac{2}{3} x$

l. $\dfrac{-x^3}{(-x)^2} = \dfrac{-x^3}{x^2} = -x^{3-2} = -x$

m. $(a^2 b^3)^3$
$= a^2 b^3 \cdot a^2 b^3 \cdot a^2 b^3$
$= a \cdot a \cdot b \cdot b \cdot b \cdot a \cdot a \cdot b \cdot b \cdot b \cdot a \cdot a \cdot b \cdot b \cdot b$
$= a^6 b^9$

n. $\left(\dfrac{a^3}{2b^2}\right)^3 = \dfrac{a^3}{2b^2} \cdot \dfrac{a^3}{2b^2} \cdot \dfrac{a^3}{2b^2}$
$= \dfrac{a \cdot a \cdot a \cdot a \cdot a \cdot a \cdot a \cdot a \cdot a}{2 \cdot 2 \cdot 2 \cdot b \cdot b \cdot b \cdot b \cdot b \cdot b}$
$= \dfrac{a^9}{8b^6}$

o. $\dfrac{3x - 9}{3} = \dfrac{3x}{3} - \dfrac{9}{3} = x - 3$

p. $\dfrac{x^2 + 2x}{x} = \dfrac{x^2}{x} + \dfrac{2x}{x} = x + 2$

6. a. $3x + 2y - 2x - 3y + 4x - 4y =$
$3x - 2x + 4x + 2y - 3y - 4y =$
$5x - 5y$

b. $x^3 - 3x^2 + x - 2x^2 + 6x - 2 =$
$x^3 - 3x^2 - 2x^2 + x + 6x - 2 =$
$x^3 - 5x^2 + 7x - 2$

c. $2(x - 2) + 3(x - 1) =$
$2x - 4 + 3x - 3 = 5x - 7$

d. $12(x - 1) - 5(x - 1) =$
$7(x - 1) = 7x - 7$

e. $3a^2 + ab + 6b^2$

7. a. $3(2x + 9y) = 6x + 27y$

b. $3(2a + 9b) = 6a + 27b$

c. $2(3x^2 + 4x - 2) = 6x^2 + 8x - 4$

d. $ab(b - ab + a^2) =$
$ab^2 - a^2 b^2 + a^3 b$

8. No. $(-x)^2$ is $(-x)(-x)$ while $-x^2$ is the opposite of $x \cdot x$.

9. a. $\pi(2^2) \approx 3.14(4) \approx 12.56 \text{ ft}^2$

b. $\pi(20^2) \approx 3.14(400) \approx 1256 \text{ ft}^2$

c. 100 times as large; 100

10. $5,616,000 \text{ sec} \cdot \dfrac{1 \text{ min}}{60 \text{ sec}} \cdot \dfrac{1 \text{ hr}}{60 \text{ min}} \cdot \dfrac{1 \text{ day}}{24 \text{ hr}}$
$= 65 \text{ days}$

11. a. $18,446,400 \text{ in}^2 \cdot \left(\dfrac{1 \text{ ft}}{12 \text{ in}}\right)^2 = 128,100 \text{ ft}^2$

b. $35 \text{ yd} \cdot \dfrac{3 \text{ ft}}{1 \text{ yd}} = 105 \text{ ft}$

$\dfrac{128,100 \text{ ft}^2}{105 \text{ ft}} = 1220 \text{ ft}$

c. $2(105 \text{ ft}) + 2(1220 \text{ ft}) = 2650 \text{ ft}$

$2650 \text{ ft} \cdot \dfrac{1 \text{ yd}}{3 \text{ ft}} = 883\tfrac{1}{3} \text{ yd}$

12. a. $2(1.5 \text{ ft}) + 2(0.5\pi \text{ ft}) = (3.0 + \pi) \text{ ft}$
$\approx 6.14 \text{ ft}$

b. $10(13 \text{ m}) = 130 \text{ m}$

13. a. $S = 2(18 \text{ ft})(6 \text{ ft}) + 2(18 \text{ ft})(8 \text{ ft}) + 2(6 \text{ ft})(8 \text{ ft}) = 600 \text{ ft}^2$

$V = (18 \text{ ft})(6 \text{ ft})(8 \text{ ft}) = 864 \text{ ft}^3$

b. $S = 2\pi(6 \text{ in})(10 \text{ in}) + 2\pi\left(6\right)^2$
$\approx 602.88 \text{ in}^2$
$V = \pi(6 \text{ in})^2(10 \text{ in}) \approx 1130.4 \text{ in}^3$

14.

	Inequality	Interval	Line Graph
0 to 20	$0 \le x \le 20$	[0, 20]	
more than 20 & less than 50	$20 < x < 50$	(20, 50)	
50 or greater	$x \ge 50$	[50, +∞)	

15.

x	y	$x + y$	$x - y$	xy	$x \div y$
-8	4	-8 + 4 = -4	-8 - 4 = -12	(-8)(4) = -32	-8 ÷ 4 = -2
-6	-8 - (-6) = -2	-8	-6 - (-2) = -4	(-6)(-2) = 12	-6 ÷ (-2) = 3
6 *	-3	6 + (-3) = 3	9	-18	6 ÷ (-3) = -2
3 *	-6	3 + (-6) = -3	9	-18	$3 \div (-6) = -\tfrac{1}{2}$

* There are two possible combinations that satisfy x - y = 9 and (x)(y) = -18.

Cumulative Review Chapters 1-2

1. **a.**

x	$y = x^2 - 1$
-2	$(-2)^2 - 1 = 4 - 1 = 3$
-1	$(-1)^2 - 1 = 1 - 1 = 0$
0	$(0)^2 - 1 = 0 - 1 = -1$
1	$(1)^2 - 1 = 1 - 1 = 0$
2	$(2)^2 - 1 = 4 - 1 = 3$
3	$(3)^2 - 1 = 9 - 1 = 8$

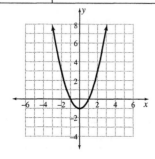

b.

x	$y = 2 - x$
-2	$2 - (-2) = 2 + 2 = 4$
-1	$2 - (-1) = 2 + 1 = 3$
0	$2 - (0) = 2$
1	$2 - (1) = 1$
2	$2 - (2) = 0$
3	$2 - (3) = -1$

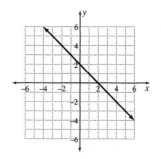

c.

x	$y = 2x + 3$
-2	$2(-2) + 3 = -4 + 3 = -1$
-1	$2(-1) + 3 = -2 + 3 = 1$
0	$2(0) + 3 = 0 + 3 = 3$
1	$2(1) + 3 = 2 + 3 = 5$
2	$2(2) + 3 = 4 + 3 = 7$
3	$2(3) + 3 = 6 + 3 = 9$

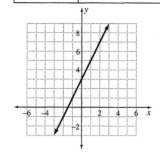

d.

x	$y = -2x$
-2	-2(-2) = 4
-1	-2(-1) = 2
0	-2(0) = 0
1	-2(1) = -2
2	-2(2) = -4
3	-2(3) = -6

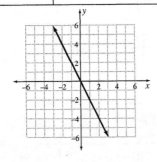

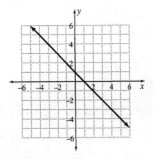

f.

x	$y = \lvert x - 1 \rvert$
-2	$\lvert -2 - 1 \rvert = \lvert -3 \rvert = 3$
-1	$\lvert -1 - 1 \rvert = \lvert -2 \rvert = 2$
0	$\lvert 0 - 1 \rvert = \lvert -1 \rvert = 1$
1	$\lvert 1 - 1 \rvert = \lvert 0 \rvert = 0$
2	$\lvert 2 - 1 \rvert = \lvert 1 \rvert = 1$
3	$\lvert 3 - 1 \rvert = \lvert 2 \rvert = 2$

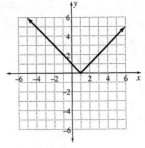

e.

x	$y = -x + 1$
-2	-(-2) + 1 = 2 + 1 = 3
-1	-(-1) + 1 = 1 + 1 = 2
0	-(0) + 1 = 0 + 1 = 1
1	-(1) + 1 = 0
2	-(2) + 1 = -1
3	-(3) + 1 = -2

3. a. 4 is less than x.

b. The product of 3 and x is subtracted from 5.

c. Opposite of the opposite of x.

d. The absolute value of the difference between 3 and x.

5. a. $3(x+1) + 5(x-2) =$
$3x + 3 + 5x - 10 = 8x - 7$

b. $8 - 2(x - 1) = 8 - 2x + 2$
$= -2x + 10$

c. $8(x + 2) - 3(x - 2) =$
$8x + 16 - 3x + 6 = 5x + 22$

d. $5 - 3(2 - x) = 5 - 6 + 3x = 3x - 1$

7.

x	y	$x + y$	$x - y$	xy	$x \div y$
-4	$6 = 2 - (-4)$	2	$-4 - 6 = -10$	$(-4)(6) = -24$	$\frac{-4}{6} = -\frac{2}{3}$
5	$-3 = -15 \div 5$	$5 + (-3) = 2$	$5 - (-3) = 8$	-15	$\frac{5}{-3} = -\frac{5}{3}$
$-\frac{2}{3} = -\frac{8}{12}$	$\frac{3}{4} = \frac{9}{12}$	$-\frac{8}{12} + \frac{9}{12} = \frac{1}{12}$	$-\frac{8}{12} - \frac{9}{12} = -\frac{17}{12}$	$-\frac{2}{3} \cdot \frac{3}{4} = -\frac{1}{2}$	$-\frac{2}{3} \div \frac{3}{4} = -\frac{2}{3} \cdot \frac{4}{3}$ $= -\frac{8}{9}$
$\frac{7}{8} = \frac{35}{40}$	$-\frac{7}{10} = -\frac{28}{40}$	$\frac{35}{40} + (-\frac{28}{40}) = \frac{7}{40}$	$\frac{35}{40} - (-\frac{28}{40}) = \frac{63}{40}$	$\frac{7}{8} \cdot (-\frac{7}{10}) = -\frac{49}{80}$	$\frac{7}{8} \div (-\frac{7}{10}) =$ $\frac{7}{8} \cdot (-\frac{10}{7}) = -\frac{5}{4}$
1.44	1.8	$1.44 + 1.8 = 3.24$	$1.44 - 1.8 = -0.36$	$1.44(1.8) = 2.592$	$1.44 \div 1.8 = 0.8$
0.25	-0.5	$0.25 + (-0.5) =$ -0.25	$0.25 - (-0.5) =$ 0.75	$0.25(-0.5) =$ -0.125	$0.25 \div (-0.5) =$ -0.5

9. a. $+68 - 74 - 26 + 32 + 14 =$
$+68 + 32 - 74 - 26 + 14 =$
$+100 - 100 + 14 = 14$

b. $-16 + 18 - 35 + 12 - 15 - 24 =$
$-16 - 24 + 18 + 12 - 35 - 15 =$
$-40 + 30 - 50 = -60$

c. $-8 - (-2) = -8 + 2 = -6$

d. $7 - (-3) = 7 + 3 = 10$

e. $-4 + (-5) = -9$

f. $-24 \div (-3) = \frac{-24}{-3} = 8$

g. $(-2)(-3)(-5) = (6)(-5) = -30$

h. $5 \div \frac{2}{3} \cdot 4 = 5 \cdot \frac{3}{2} \cdot 4 = 5 \cdot 3 \cdot 2 = 30$

i. $7 \div \frac{3}{2} \cdot 6 = 7 \cdot \frac{2}{3} \cdot 6 = 7 \cdot 2 \cdot 2 = 28$

j. $\frac{a}{b} \div \frac{a}{b} = \frac{a}{b} \cdot \frac{b}{a} = 1$

k. $\frac{x}{y} \div \frac{-x}{y} = \frac{x}{y} \cdot \frac{y}{-x} = -1$

l. $\frac{6a + 2}{3} = \frac{6a}{3} + \frac{2}{3} = 2a + \frac{2}{3}$

m. $\frac{xy - x^2}{x} = \frac{xy}{x} - \frac{x^2}{x} = y - x$

11. $(40 \text{ in})(8 \text{ in})(20 \text{ ft})\left(\dfrac{1 \text{ ft}}{12 \text{ in}}\right)^2\left(\dfrac{1 \text{ yd}}{3 \text{ ft}}\right)^3 =$

$\dfrac{6400 \text{ in}^2\text{ft} \cdot 1 \text{ ft}^2 \cdot 1 \text{ yd}^3}{144 \text{ in}^2 \cdot 27 \text{ ft}^3} \approx 1.65 \text{ yd}^3$

13. a The distance from Erie to Pittsburgh equals the distance from Pittsburgh to Erie. (for example)

    b.

	Erie	Pittsburg	Reading	Scranton
Erie	–	135	325	300
Pittsburg	135	–	255	272
Reading	325	255	–	99
Scranton	300	272	99	–

    c. The commutative properties of addition and multiplication.

15. a. $90 - (10 - 4) = 84$

    b. $80 - (20 - 4) = 64$

    c. $70 - (30 - 4) = 44$

    d. Since points are only given for correct answers, the highest score would be obtained if the remaining questions are correct.
$99 \text{ (correct)} - 1 \text{ (incorrect)} = 98$

    e. Yes. With 99 correct and 1 blank, a student would have a score of 99.

Chapter 3

1. **a.** conditional, 2 variables x, y

    b. conditional, 1 variable x

    c. conditional, 3 variables A, b, h

    d. identity

    e. identity

    f. identity

3. **a.** linear; can be put in the form $y = mx + b$.

    b. linear; can be put in the form $mx + b = 0$

    c. nonlinear; cannot be written as either $y = mx + b$ or $mx + b = 0$

    d. linear; can be put in the form $y = mx + b$.

    e. nonlinear; cannot be written as either $y = mx + b$ or $mx + b = 0$

    f. linear; can be put in the form $mx + b = 0$

5. $\dfrac{x}{2} = 8$

7. $6 + x = 4$

9. $2x - 15 = -9$

11. $19 = 3 + 2x$

13. $3x - 4 = 17$

15. $26 - 4x = 2$

17. Four less than a number is six.

19. Five less than the product of three and a number is sixteen.

21. Two thirds of a number is twenty-four.

23. Six less twice a number is ten.

25. $y = 2x + 5$

27. $y = 3x - 6$

29. $y = \dfrac{1}{2}x - 5$

31. Let x = cost of the meal, and y = tip.
    $y = 0.15x$

33. Let x = amount of wages, and y = the Medicare payment.
    $y = 0.0145x$

35. Let x = number of pounds, and y = the total cost of the candy.
    $y = 2.49x$

37. Let x = the speed, and y = total distance traveled.
    $y = 3x$

39. Let x = number of credit hours, and y = total cost of tuition and fees.
    $y = 75x + 32$

41. Let x = number of beverages, and y = remaining balance on coffee card.
    $y = 26 - 4.25x$ or $y = -4.25x + 26$

43. $3(x + 5) = 21$

45. $5(9 - x) = -35$

47. $\dfrac{1}{2}(8 + x) = y$

49. Let x = number of minutes, and y = cost of call.

    $y = 0.85x$ if $x \le 2$

    $y = 0.85(2) + 0.15(x - 2)$ if $x > 2$
    $= 1.7 + 0.15x - 0.3$
    $= 0.15x + 1.4$

51. Let x = number of transcripts ordered, and y = total cost. ($x \ge 1$)
    $y = 2(x - 1) + 5$
    $= 2x - 2 + 5$
    $= 2x + 3$

53. Let x = bowling average, and y = the handicap.
    $y = 0.8(200 - x)$ for $x \le 200$

55. a. $x = 5$ so $7 = x + 2$

    b. $x = 6$ so $5 = x - 1$ and $7 = x + 1$

    c. $x = 7$ so $5 = x - 2$

57. a. $x = 5$ so $7 = x + 2$ and $9 = x + 4$

    b. $x = 7$ so $9 = x + 2$

    c. $x = 9$ so $5 = x - 4$

59. Let $x =$ the first integer.
Consecutive integers differ by 1 so we have:
$$y = x + (x + 1) + (x + 2)$$

61. Let $x =$ the first even integer.
Consecutive even integers differ by 2 so we have:
$$y = x + (x + 2) + (x + 4) + (x + 6)$$

63. a.

	$x = 11$	$x = 12$
x	11	12
$x + 1$	$11 + 1 = 12$	$12 + 1 = 13$
$x + 3$	$11 + 3 = 14$	$12 + 3 = 15$

    b.

	$x = 11$	$x = 12$
x	11	12
$x + 2$	$11 + 2 = 13$	$12 + 2 = 14$
$x + 4$	$11 + 4 = 15$	$12 + 4 = 16$

    c.   part (b)

    d.   Even numbers and odd numbers alternate. The numbers are two units apart in both sets.

    e.   The numbers are two units apart in both sets.

    f.

	$x = 11$	$x = 12$
$x - 2$	$11 - 2 = 9$	$12 - 2 = 10$
x	11	12
$x + 2$	$11 + 2 = 13$	$12 + 2 = 14$

	$x = 11$	$x = 12$
$x - 4$	$11 - 4 = 7$	$12 - 4 = 8$
$x - 2$	$11 - 2 = 9$	$12 - 2 = 10$
x	11	12

65. An identity equation is true for all values of the variable(s) while a conditional equation is true for only certain values.

67. A linear equation can be written in the form $y = mx + b$ (two variables) or $mx + b = 0$ (one variable) where $m \neq 0$.

69. Answers may vary. Some possible phrases: *sum, difference, the quantity*

71. Even integers are two units apart and odd integers are also two units apart. Their descriptions are the same.

73. a. $\dfrac{28}{35} = \dfrac{7 \cdot 4}{7 \cdot 5} = \dfrac{4}{5}$

    b. $0.8 = \dfrac{8}{10} = \dfrac{8 \cdot 10}{10 \cdot 10} = \dfrac{80}{100} = 80\%$

    c. $\dfrac{45}{6} = \dfrac{42 + 3}{6} = \dfrac{42}{6} + \dfrac{3}{6} = 7 + \dfrac{1}{2} = 7\dfrac{1}{2}$

    d. $\dfrac{32}{40} = \dfrac{8 \cdot 4}{8 \cdot 5} = \dfrac{4}{5}$

    e. $0.3 = \dfrac{3}{10} = \dfrac{3 \cdot 10}{10 \cdot 10} = \dfrac{30}{100} = 30\%$

    f. $4 - 2(3 - 6) = 4 - 2(-3)$
$$= 4 + 6$$
$$= 10$$

    g. $8 - 3(2 - 8) = 8 - 3(-6)$
$$= 8 + 18$$
$$= 26$$

    h. $\dfrac{24}{9} = \dfrac{18 + 6}{9} = \dfrac{18}{9} + \dfrac{6}{9} = 2 + \dfrac{2}{3} = 2\dfrac{2}{3}$

Exercises 3.2

1. a. $2(4) = 8$; Correct solution

    b. $(-6) + 1 = -5$; Correct solution

    c. $(12) - 3 = 15$; Incorrect, 3 was subtracted from 15, need to add 3.

3. a. $55(2) = 440$; Incorrect, 440 was divided by 22, it should be divided by 55.

b. $\frac{1}{2}(5) = 10$; Incorrect, 10 was divided by 2, need to multiply by 2.

c. $1.05(40) = 42$; Correct solution

5. The first equation was divided by 2 on both sides.

7. 5 was subtracted from both sides of the first equation.

9. 4 was added to both sides of the first equation.

11. Both sides of the first equation were multiplied by 2.

13. Add 5 to both sides.
$$x - 5 = 8$$
$$x - 5 + 5 = 8 + 5$$
$$x = 13$$

Check: $13 - 5 \overset{?}{=} 8$ ✓

15. Subtract 12 from both sides.
$$9 = x + 12$$
$$9 - 12 = x + 12 - 12$$
$$-3 = x \text{ or } x = -3$$

Check: $9 \overset{?}{=} -3 + 12$ ✓

17. Add 6 to both sides.
$$x - 6 = -10$$
$$x - 6 + 6 = -10 + 6$$
$$x = -4$$

Check: $-4 - 6 \overset{?}{=} -10$ ✓

19. Divide both sides by 2.
$$2x = 26$$
$$\frac{2x}{2} = \frac{26}{2}$$
$$x = 13$$

Check: $2(13) \overset{?}{=} 26$ ✓

21. Divide both sides by 8.
$$3 = 8x$$
$$\frac{3}{8} = \frac{8x}{8}$$
$$\frac{3}{8} = x \text{ or } x = \frac{3}{8}$$

Check: $3 \overset{?}{=} 8\left(\frac{3}{8}\right)$ ✓

23. Multiply both sides by 4.
$$\frac{x}{4} = 16$$
$$4 \cdot \frac{x}{4} = 4 \cdot 16$$
$$x = 64$$

Check: $\frac{64}{4} \overset{?}{=} 16$ ✓

25. Multiply both sides by -12.
$$\frac{-x}{12} = 4$$
$$-12 \cdot \frac{-x}{12} = -12 \cdot 4$$
$$x = -48$$

Check: $\frac{-(-48)}{12} \overset{?}{=} 4$ ✓

27. Multiply both sides by 2.
$$-4 = \frac{1}{2}x$$
$$2 \cdot (-4) = 2 \cdot \frac{1}{2}x$$
$$-8 = x \text{ or } x = -8$$

Check: $-4 \overset{?}{=} \frac{1}{2}(-8)$ ✓

29. Multiply both sides by $-\dfrac{4}{3}$.

$$-\frac{3}{4}x = 12$$

$$\left(-\frac{4}{3}\right)\left(-\frac{3}{4}x\right) = \left(-\frac{4}{3}\right)(12)$$

$$x = -\frac{48}{3}$$

$$x = -16$$

Check: $-\dfrac{3}{4}(-16) \overset{?}{=} 12$ ✓

31. Multiply both sides by $\dfrac{4}{3}$.

$$\frac{3}{4}x = 18$$

$$\frac{4}{3}\left(\frac{3}{4}\right)x = \frac{4}{3}\cdot 18$$

$$x = \frac{72}{3}$$

$$x = 24$$

Check: $\dfrac{3}{4}(24) \overset{?}{=} 18$ ✓

33. Subtract $\dfrac{2}{3}$ from both sides.

$$\frac{2}{3} + x = 16$$

$$\frac{2}{3} + x - \frac{2}{3} = 16 - \frac{2}{3}$$

$$x = \frac{48}{3} - \frac{2}{3}$$

$$x = \frac{46}{3} \text{ or } 15\frac{1}{3}$$

Check: $\dfrac{2}{3} + \dfrac{46}{3} \overset{?}{=} 16$ ✓

35. Subtract $\dfrac{3}{4}$ from both sides, then multiply both sides by -1.

$$\frac{3}{4} - x = 18$$

$$\frac{3}{4} - x - \frac{3}{4} = 18 - \frac{3}{4}$$

$$-x = \frac{72}{4} - \frac{3}{4}$$

$$-x = \frac{69}{4}$$

$$(-1)(-x) = (-1)\left(\frac{69}{4}\right)$$

$$x = -\frac{69}{4} \quad \text{or} \quad x = -17\frac{1}{4}$$

Check: $\dfrac{3}{4} - \left(-\dfrac{69}{4}\right) \overset{?}{=} 18$ ✓

37. x times -2 plus 4 gives 3

$$4 - 2x = 3$$

$$4 - 2x - 4 = 3 - 4$$

$$-2x = -1$$

$$\frac{-2x}{-2} = \frac{-1}{-2}$$

$$x = \frac{1}{2}$$

Check: $4 - 2\left(\dfrac{1}{2}\right) \overset{?}{=} 3$ ✓

39. x times $\dfrac{1}{2}$ subtract 4 gives 3

$$\frac{1}{2}x - 4 = 3$$

$$\frac{1}{2}x - 4 + 4 = 3 + 4$$

$$\frac{1}{2}x = 7$$

$$2\cdot\frac{1}{2}x = 2\cdot 7$$

$$x = 14$$

Check: $\dfrac{1}{2}(14) - 4 \overset{?}{=} 3$ ✓

41. x times $\frac{1}{4}$ subtract 2 gives 3

$$\frac{1}{4}x - 2 = 3$$
$$\frac{1}{4}x - 2 + 2 = 3 + 2$$
$$\frac{1}{4}x = 5$$
$$4 \cdot \frac{1}{4}x = 4 \cdot 5$$
$$x = 20$$

Check: $\frac{1}{4}(20) - 2 \overset{?}{=} 3$ ✓

43.
$$4x + 3 = 23$$
$$4x + 3 - 3 = 23 - 3$$
$$4x = 20$$
$$\frac{4x}{4} = \frac{20}{4}$$
$$x = 5$$

Check: $4(5) + 3 \overset{?}{=} 23$ ✓

45.
$$3x - 2 = 43$$
$$3x - 2 + 2 = 43 + 2$$
$$3x = 45$$
$$\frac{3x}{3} = \frac{45}{3}$$
$$x = 15$$

Check: $3(15) - 2 \overset{?}{=} 43$ ✓

47.
$$10 - x = -2$$
$$10 - x - 10 = -2 - 10$$
$$-x = -12$$
$$\frac{-x}{-1} = \frac{-12}{-1}$$
$$x = 12$$

Check: $10 - 12 \overset{?}{=} -2$ ✓

49.
$$3 = 10 - x$$
$$3 - 10 = 10 - x - 10$$
$$-7 = -x$$
$$\frac{-7}{-1} = \frac{-x}{-1}$$
$$7 = x \text{ or } x = 7$$

Check: $3 \overset{?}{=} 10 - 7$ ✓

51.
$$0 = 3 - 3x$$
$$0 - 3 = 3 - 3x - 3$$
$$-3 = -3x$$
$$\frac{-3}{-3} = \frac{-3x}{-3}$$
$$1 = x \text{ or } x = 1$$

Check: $0 \overset{?}{=} 3 - 3(1)$ ✓

53.
$$3 - 3x = 9$$
$$3 - 3x - 3 = 9 - 3$$
$$-3x = 6$$
$$\frac{-3x}{-3} = \frac{6}{-3}$$
$$x = -2$$

Check: $3 - 3(-2) \overset{?}{=} 9$ ✓

55.
$$\frac{1}{2}x + 3 = -2$$
$$\frac{1}{2}x + 3 - 3 = -2 - 3$$
$$\frac{1}{2}x = -5$$
$$2 \cdot \frac{1}{2}x = 2(-5)$$
$$x = -10$$

Check: $\frac{1}{2}(-10) + 3 \overset{?}{=} -2$ ✓

57.
$$0 = \frac{3}{2}x + 3$$
$$0 - 3 = \frac{3}{2}x + 3 - 3$$
$$-3 = \frac{3}{2}x$$
$$\left(\frac{2}{3}\right)(-3) = \frac{2}{3} \cdot \frac{3}{2}x$$
$$-2 = x \ \text{ or } \ x = -2$$

Check: $0 \overset{?}{=} \frac{3}{2}(-2) + 3 \ \checkmark$

59.
$$-4 = \frac{2}{3}x - 2$$
$$-4 + 2 = \frac{2}{3}x - 2 + 2$$
$$-2 = \frac{2}{3}x$$
$$\left(\frac{3}{2}\right)(-2) = \frac{3}{2} \cdot \frac{2}{3}x$$
$$-3 = x \ \text{ or } \ x = -3$$

Check: $-4 \overset{?}{=} \frac{2}{3}(-3) - 2$

61.
$$4.2x - 3 = -9.3$$
$$4.2x - 3 + 3 = -9.3 + 3$$
$$4.2x = -6.3$$
$$\frac{4.2x}{4.2} = \frac{-6.3}{4.2}$$
$$x = -1.5$$

Check: $4.2(-1.5) - 3 \overset{?}{=} -9.3 \ \checkmark$

63.
$$7.5 = 4.2x - 3$$
$$7.5 + 3 = 4.2x - 3 + 3$$
$$10.5 = 4.2x$$
$$\frac{10.5}{4.2} = \frac{4.2x}{4.2}$$
$$2.5 = x \ \text{ or } \ x = 2.5$$

Check: $7.5 \overset{?}{=} 4.2(2.5) - 3 \ \checkmark$

65.
$$0.03x = 9$$
$$\frac{0.03x}{0.03} = \frac{9}{0.03}$$
$$x = 300$$
The cash advance is $300.

67.
$$0.03x = 31.5$$
$$\frac{0.03x}{0.03} = \frac{31.5}{0.03}$$
$$x = 10,500$$
The cash advance is $10,500.

69. Let x = the first integer.
$$x + (x+1) + (x+2) = 42$$
$$3x + 3 = 42$$
$$3x + 3 - 3 = 42 - 3$$
$$3x = 39$$
$$\frac{3x}{3} = \frac{39}{3}$$
$$x = 13$$
The three integers are 13, 14, and 15.

71. Let x = the first odd integer.
$$x + (x+2) + (x+4) = 177$$
$$3x + 6 = 177$$
$$3x + 6 - 6 = 177 - 6$$
$$3x = 171$$
$$\frac{3x}{3} = \frac{171}{3}$$
$$x = 57$$
The three odd integers are 57, 59, and 61.

73. Let x = the first publication year.
$$x + (x+7) + (x+14) + (x+21)$$
$$+ (x+28) + (x+35) + (x+42)$$
$$= 13,741$$
$$7x + 147 = 13,741$$
$$7x + 147 - 147 = 13,741 - 147$$
$$7x = 13,594$$
$$\frac{7x}{7} = \frac{13,594}{7}$$
$$x = 1942$$
The first book was published in 1942.

75. Subtraction is the same as adding the opposite.

77. Add a to both sides and simplify the right side.

79. This gives a positive sign on the *x* so there is no need to multiply or divide both sides by a negative number. The next step is to subtract 8.

81. 7 is added twice on the left side instead of once on each side.

83. $110 = 55t$

$$\frac{110}{55} = \frac{55t}{55}$$

$$2 = t$$

85.
$$212 = \frac{9}{5}C + 32$$

$$212 - 32 = \frac{9}{5}C + 32 - 32$$

$$180 = \frac{9}{5}C$$

$$\frac{5}{9} \cdot 180 = \frac{5}{9} \cdot \frac{9}{5}C$$

$$100 = C$$

87. $D = 55t$

$$200 = 55t$$

$$\frac{200}{55} = \frac{55t}{55}$$

$$\frac{40}{11} = t \text{ or } t = 3\frac{7}{11}$$

89. $D = 3r$

$$200 = 3r$$

$$\frac{200}{3} = \frac{3r}{3}$$

$$66\frac{2}{3} = r$$

91. $T = 0.06p$

$$0.1 = 0.06p$$

$$\frac{0.1}{0.06} = \frac{0.06p}{0.06}$$

$$\frac{10}{6} = p \text{ or } p \approx \$1.67$$

93.
$$C = 5.50 + 0.0821x$$

$$75 = 5.50 + 0.0821x$$

$$75 - 5.50 = 5.50 + 0.0821x - 5.50$$

$$69.50 = 0.0821x$$

$$\frac{69.50}{0.0821} = \frac{0.0821x}{0.0821}$$

$$846.53 \approx x \text{ or } x \approx 846.53 \text{ kwh}$$

95.
$$A = \frac{1}{2}bh$$

$$15 = \frac{1}{2}(3)h$$

$$15 = \frac{3}{2}h$$

$$\frac{2}{3} \cdot 15 = \frac{2}{3} \cdot \frac{3}{2}h$$

$$10 = h \text{ or } h = 10$$

Exercises 3.3

1. a. From the table, when $y = 8$, $x = 2$.

b. From the table, when $y = 6$, $x = 4$.

c. For each unit decrease in *y*, there is a unit increase in *x*.
Extending the table, when $y = 3$, $x = 7$.

d. For each unit decrease in *y*, there is a unit increase in *x*.
Extending the table, when $y = -2$, $x = 12$.

3. a. From the table, when $y = -2$, $x = 1$.

b. From the table, when $y = -6$, $x = 2$.

c. For every 4 unit increase in *y*, there is a unit decrease in *x*.
Extending the table, when $y = 6$, $x = -1$.

d. For every 4 unit decrease in *y*, there is a unit increase in *x*. Therefore, for every two unit decrease in *y*, there is a $\frac{1}{2}$ unit increase in *x*.
Extending the table, when $y = -8$,
$$x = 2\frac{1}{2}.$$

5. For each unit increase in y, the value of x increases by $\dfrac{1}{2} = 0.5$ units.

x	$y = 2x+1$
3	7
3.5	8
4	9
4.5	10
5	11
5.5	12

7. For each unit increase in y, the value of x increases by $\dfrac{1}{5} = 0.2$ units.

x	$y = 5x-4$
3	11
3.2	12
3.4	13
3.6	14
3.8	15
4	16

9. a. Find the intersection of the graphs of $y = 1$ and $y = \dfrac{1}{2}x+3$.

$\dfrac{1}{2}x+3 = 1$ when $x = -4$.

b. Find the intersection of the graphs of $y = 0$ and $y = \dfrac{1}{2}x+3$.

$\dfrac{1}{2}x+3 = 0$ when $x = -6$.

c. Find the intersection of the graphs of $y = -2$ and $y = \dfrac{1}{2}x+3$.

$\dfrac{1}{2}x+3 = -2$ when $x = -10$.

11. a. Find the intersection of the graphs of $y = 1$ and $y = 4x-3$.
$4x-3 = 1$ when $x = 1$.

b. Find the intersection of the graphs of $y = -7$ and $y = 4x-3$.
$4x-3 = -7$ when $x = -1$.

c. Find the intersection of the graphs of $y = 5$ and $y = 4x-3$.
$4x-3 = 5$ when $x = 2$.

13. a. Find the intersection of the graphs of $y = -3$ and $y = -2x^2+5$.
$x = \{-2, 2\}$

b. Find the intersection of the graphs of $y = 3$ and $y = -2x^2+5$.
$x = \{-1, 1\}$

c. Find the intersection of the graphs of $y = 5$ and $y = -2x^2+5$.
$x = \{0\}$

d. Find the intersection of the graphs of $y = 6$ and $y = -2x^2+5$. $x = \{\ \}$ (the graphs never cross)

15. a.

x	$y = 8-3(x+2)$
-3	$8-3(-3+2) = 11$
-2	$8-3(-2+2) = 8$
-1	$8-3(-1+2) = 5$
0	$8-3(0+2) = 2$
1	$8-3(1+2) = -1$
2	$8-3(2+2) = -4$
3	$8-3(3+2) = -7$

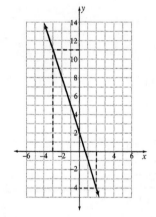

b.
$$8 - 3(x+2) = 11$$
$$8 - 3x - 6 = 11$$
$$-3x + 2 = 11$$
$$-3x + 2 - 2 = 11 - 2$$
$$-3x = 9$$
$$\frac{-3x}{-3} = \frac{9}{-3}$$
$$x = -3$$

This agrees with the table.

c.
$$8 - 3(x+2) = 0$$
$$8 - 3x - 6 = 0$$
$$-3x + 2 = 0$$
$$-3x + 2 - 2 = 0 - 2$$
$$-3x = -2$$
$$\frac{-3x}{-3} = \frac{-2}{-3}$$
$$x = \frac{2}{3}$$

d.
$$8 - 3(x+2) = -4$$
$$8 - 3x - 6 = -4$$
$$-3x + 2 = -4$$
$$-3x + 2 - 2 = -4 - 2$$
$$-3x = -6$$
$$\frac{-3x}{-3} = \frac{-6}{-3}$$
$$x = 2$$

17.
$$2(x-2) = 4$$
$$2x - 4 = 4$$
$$2x - 4 + 4 = 4 + 4$$
$$2x = 8$$
$$\frac{2x}{2} = \frac{8}{2}$$
$$x = 4$$

19.
$$5(x+4) = 12$$
$$5x + 20 = 12$$
$$5x + 20 - 20 = 12 - 20$$
$$5x = -8$$
$$\frac{5x}{5} = \frac{-8}{5}$$
$$x = -1.6$$

21.
$$2(2-x) = -5$$
$$4 - 2x = -5$$
$$4 - 2x - 4 = -5 - 4$$
$$-2x = -9$$
$$\frac{-2x}{-2} = \frac{-9}{-2}$$
$$x = 4.5$$

23.
$$3(x+2) = -7$$
$$3x + 6 = -7$$
$$3x + 6 - 6 = -7 - 6$$
$$3x = -13$$
$$\frac{3x}{3} = \frac{-13}{3}$$
$$x = -4\frac{1}{3}$$

25.
$$\frac{1}{2}(6 - 3x) = 9$$
$$3 - \frac{3}{2}x = 9$$
$$3 - \frac{3}{2}x - 3 = 9 - 3$$
$$-\frac{3}{2}x = 6$$
$$\left(-\frac{2}{3}\right)\left(-\frac{3}{2}\right)x = \left(-\frac{2}{3}\right) \cdot 6$$
$$x = -4$$

27.
$$\frac{2}{3}(x-5) = -2$$
$$\frac{2}{3}x - \frac{10}{3} = -2$$
$$\frac{2}{3}x - \frac{10}{3} + \frac{10}{3} = -2 + \frac{10}{3}$$
$$\frac{2}{3}x = \frac{4}{3}$$
$$\frac{3}{2} \cdot \frac{2}{3}x = \frac{3}{2} \cdot \frac{4}{3}$$
$$x = 2$$

29. $\quad 4-3(x+1)=16$

$\qquad 4-3x-3=16$

$\qquad\qquad -3x+1=16$

$\qquad -3x+1-1=16-1$

$\qquad\qquad\quad -3x=15$

$\qquad\qquad \dfrac{-3x}{-3}=\dfrac{15}{-3}$

$\qquad\qquad\qquad x=-5$

31. $\quad 7-3(4-x)=-26$

$\qquad 7-12+3x=-26$

$\qquad\qquad 3x-5=-26$

$\qquad 3x-5+5=-26+5$

$\qquad\qquad\quad 3x=-21$

$\qquad\qquad \dfrac{3x}{3}=\dfrac{-21}{3}$

$\qquad\qquad\quad x=-7$

33. $\quad 8-6(x-7)=74$

$\qquad 8-6x+42=74$

$\qquad\quad -6x+50=74$

$\quad -6x+50-50=74-50$

$\qquad\qquad\quad -6x=24$

$\qquad\qquad \dfrac{-6x}{-6}=\dfrac{24}{-6}$

$\qquad\qquad\qquad x=-4$

35. Let x be the number.

$\qquad 2(x+5)=14$

$\qquad 2x+10=14$

$\quad 2x+10-10=14-10$

$\qquad\qquad 2x=4$

$\qquad\qquad \dfrac{2x}{2}=\dfrac{4}{2}$

$\qquad\qquad x=2$

37. Let x be the number.

$\qquad -2(x-4)=6$

$\qquad -2x+8=6$

$\quad -2x+8-8=6-8$

$\qquad\qquad -2x=-2$

$\qquad\qquad \dfrac{-2x}{-2}=\dfrac{-2}{-2}$

$\qquad\qquad x=1$

39. Let x be the number.

$\qquad \dfrac{1}{2}(x-5)=24$

$\qquad \dfrac{1}{2}x-\dfrac{5}{2}=24$

$\qquad \dfrac{1}{2}x-\dfrac{5}{2}+\dfrac{5}{2}=24+\dfrac{5}{2}$

$\qquad\qquad \dfrac{1}{2}x=\dfrac{53}{2}$

$\qquad 2\cdot\dfrac{1}{2}x=2\cdot\dfrac{53}{2}$

$\qquad\qquad x=53$

41. Let x be the number of miles driven.

$\qquad 65+0.15(x-100)=100$

$\qquad 65+0.15x-15=100$

$\qquad\quad 0.15x+50=100$

$\quad 0.15x+50-50=100-50$

$\qquad\qquad 0.15x=50$

$\qquad\qquad \dfrac{0.15x}{0.15}=\dfrac{50}{0.15}$

$\qquad\qquad x\approx 333.33$

Roughly 333.333 miles can be driven for a total of $100.

43. Let x be the number of copies made.

$\qquad 2.00+1.50(x-1)=6.50$

$\qquad 2.00+1.50x-1.50=6.50$

$\qquad\quad 1.50x+0.50=6.50$

$\quad 1.50x+0.50-0.50=6.50-0.50$

$\qquad\qquad 1.50x=6.00$

$\qquad\qquad \dfrac{1.50x}{1.50}=\dfrac{6.00}{1.50}$

$\qquad\qquad x=4$

A total of 4 copies can be made for $6.50.

45. Let x be the number of children invited.

$\qquad 26+2.5(x-8)=70$

$\qquad 26+2.5x-20=70$

$\qquad\quad 2.5x+6=70$

$\qquad 2.5x+6-6=70-6$

$\qquad\qquad 2.5x=64$

$\qquad\qquad \dfrac{2.5x}{2.5}=\dfrac{64}{2.5}$

$\qquad x=25.6$

Up to 25 children can be invited on a $70 budget.

47. Find the value for n in the output column. Then find the corresponding value from the input column.

49. $\{\ \}$ represents the empty set and contains no members.
$\{0\}$ represents a set whose only member is 0.

51. True

53. Don't multiply the right side of the equation by 2.

Mid-Chapter 3 Test

1. a. conditional

b. identity

c. identity

d. conditional

2. a. Equivalent, 5 was subtracted from both sides of the first equation to obtain the second.

b. Not equivalent, half of 8 is not 16

c. Not equivalent, 2x + 3x is 5x

d. Not equivalent, $3x-2=8$ is equivalent to $3x=10$.

3. 3(-4) - 4 = -16?
-12 - 4 = -16, true statement

4. 4(-3) - 3 = -9?
-12 - 3 = -9, false statement
x = -3 is not a solution

5. $-8(\frac{1}{2})+5=1$?
-4 + 5 = 1, true statement

6. $-6 - 10(\frac{1}{2}) = -11$?
-6 - 5 = -11, true statement

7. x - 4 = 3, x = 3 + 4, x = 7

8. $\frac{2}{3}x=24$, $x=\frac{24\cdot3}{2}$, $x=36$

9. $2x+3=-7$, $2x=-10$
$x=\frac{-10}{2}$, $x=-5$

10. $\frac{1}{2}x-8=-1$, $\frac{1}{2}x=7$
$x=7(2)$, $x=14$

11. $3x=\frac{1}{2}$, $x=\frac{1}{2}\cdot\frac{1}{3}$, $x=\frac{1}{6}$

12. 3 - 2x = 8, -2x = 5, $x=\frac{5}{-2}$, $x=-2\frac{1}{2}$

13. $2x+5=10$

14. Let x be the first even number.
$x+(x+2)+(x+4)=126$

15. Let x be the number of credit hours.
$\$545=\$85x+\$35$

16. Let x be the number of minutes used.
$\$7.20=\$20-\$0.40x$

17. Five is 4 less than three times a number.

18. Four more than twice a number is -3.

19. a. From the table, when $y=5$, $x=2$.

b. From the table, when $y=8$, $x=8$.

c. For each unit decrease in y, there is a 2 unit decrease in x.
Extending the table, when $y=4$, $x=0$.

d. Extending the table, when $y=6.5$, $x=5$

20. a. From the graph, when y = -5, $x=2$.

b. From the graph, when y = 1, x = -4.

c. From the graph, when y = -2, x = -1.

21. a. From the graph, when y = 6, x = {-4, 3}

b. From the graph, when y = 0, x = {-3, 2}

c. From the graph, when y = -7, x = {   }

22. There should be no equal sign between 10 and x. A comma would be correct.

23. a. Let x be the number of children and y be the total cost.

$$y = \begin{cases} 85 & \text{for } x \le 11 \\ 85 + 5(x-11) & \text{for } x > 11 \end{cases}$$

x (children)	y (total $)
2 to 10	85
12	90
14	100
16	110
18	120
20	130

b. $85 (from table)

c $130 (from table)

d.
$$100 = 5(x - 11) + 85$$
$$100 = 5x - 55 + 85$$
$$100 = 5x + 30$$
$$100 - 30 = 5x + 30 - 30$$
$$70 = 5x$$
$$\frac{70}{5} = \frac{5x}{5}$$
$$14 = x \quad \text{or} \quad x = 14$$

For $100, 14 children could be at the party.

Exercises 3.4

1.

x	$5x - 8$	$2(x + 2)$
-1	5(-1) - 8 = -13	2(-1 + 2) = 2
0	5(0) - 8 = -8	2(0 + 2) = 4
1	5(1) - 8 = -3	2(1 + 2) = 6
2	5(2) - 8 = 2	2(2 + 2) = 8
3	5(3) - 8 = 7	2(3 + 2) = 10
4	**5(4) - 8 = 12**	**2(4 + 2) = 12**

Common ordered pair (4, 12), $x = 4$

3.

x	$3(x - 3)$	$6(x - 2)$
-1	3(-1 - 3) = -12	6(-1 - 2) = -18
0	3(0 - 3) = -9	6(0 - 2) = -12
1	**3(1 - 3) = -6**	**6(1 - 2) = -6**
2	3(2 - 3) = -3	6(2 - 2) = 0
3	3(3 - 3) = 0	6(3 - 2) = 6
4	3(4 - 3) = 3	6(4 - 2) = 12

Common ordered pair $(1, -6)$, $x = 1$

5. a. $3(2 - x) = 4 - x$ when $x = 1$ (intersection of 2 lines)

b. $3(2 - x) = 6$ when $x = 0$ (y-intercept)

c. $4 - x = 6$ when $x = -2$

d. $4 - x = 2$ when $x = 2$

e. $3(2 - x) = 0$ when $x = 2$ (x-intercept)

7. a. $1 - 3x = 2(x - 2)$ when $x = 1$ (intersection of 2 lines)

b. $2(x - 2) = 2$ when $x = 3$

c. $1 - 3x = 4$ when $x = -1$

d. $1 - 3x = -5$ when $x = 2$

e. $2(x - 2) = 0$ when $x = 2$ (x-intercept)

9.
$$2(x - 3) = 0$$
$$2x - 6 = 0$$
$$2x = 6$$
$$x = 3$$

11.
$$-2(x + 1) = 5$$
$$-2x - 2 = 5$$
$$-2x = 7$$
$$x = -\frac{7}{2} \quad \text{or} \quad x = -3\frac{1}{2}$$

13. $2(3x+1) = 5 - 3x$

$6x + 2 = 5 - 3x$

$9x + 2 = 5$

$9x = 3$

$x = \dfrac{1}{3}$

15. $4x + 6 = 2(1 + 3x) + 1$

$4x + 6 = 2 + 6x + 1$

$4x + 6 = 6x + 3$

$-2x + 6 = 3$

$-2x = -3$

$x = \dfrac{3}{2}$ or $x = 1\dfrac{1}{2}$

17. $7(x+1) = 11 + x$

$7x + 7 = 11 + x$

$6x + 7 = 11$

$6x = 4$

$x = \dfrac{2}{3}$

19. $2(x+3) = 3x - 2$

$2x + 6 = 3x - 2$

$-x + 6 = -2$

$-x = -8$

$x = 8$

21. $\frac{3}{5}x = 15$

$x = \dfrac{15 \cdot 5}{3}, \quad x = 25$

23. $\frac{1}{3}x = x - 12$

$-\frac{2}{3}x = -12, \quad x = 18$

25. $\frac{3}{4}x = x + 3$

$-\frac{1}{4}x = 3, \quad x = -12$

27. $\frac{1}{2}(x+5) = x - 4$

$x + 5 = 2x - 8$

$x = 13$

29. $6 - 4(x - 2) = 22$

$6 - 4x + 8 = 22, \quad 14 - 4x = 22$

$-4x = 8, \quad x = -2$

31. $2(8 - x) = 1 + 4x$

$16 - 2x = 1 + 4x$

$15 = 6x, \quad x = \frac{15}{6}, \quad x = 2\frac{1}{2}$

33. $2 - 5(x + 1) = 4 - 3x$

$2 - 5x - 5 = 4 - 3x,$

$-5x - 3 = 4 - 3x$

$-2x = 7, \quad x = -\frac{7}{2}, \quad x = -3\frac{1}{2}$

35. $7 - 2(x - 1) = 5 - 3x$

$7 - 2x + 2 = 5 - 3x$

$9 - 2x = 5 - 3x$

$x = -4$

37. $5 - 2(x - 3) = 3(x - 3)$

$5 - 2x + 6 = 3x - 9$

$11 - 2x = 3x - 9$

$11 + 9 = 3x + 2x$

$20 = 5x, \quad x = 4$

39. Distribute the 2; Distributive property

41. Add -7 to both sides; Addition property

43. Add x to both sides; Addition property

45. a.   <u>Sense Rent-a-Car:</u>

$y = 70 + 0.10x$

<u>Herr's Rent-a-Car:</u>

$y = 40 + 0.40x$

    b. $70 + 0.10x = 40 + 0.40x$

$70 + 0.10x - 70 = 40 + 0.40x - 70$

$0.10x = 0.40x - 30$

$0.10x - 0.40x = 0.40x - 30 - 0.40x$

$-0.30x = -30$

$\dfrac{-0.30x}{-0.30} = \dfrac{-30}{-0.30}$

$x = 100$

The costs are the same at 100 miles.

    c. $y = 70 + 0.10x$

$y = 70 + 0.10(100)$

$y = 70 + 10$

$y = 80$

The ordered pair is $(100, 80)$. The cost for each is \$80 at 100 miles.

    d.   Herr's graph would be steeper because it has a higher cost per mile (coefficient on x).

e. After $x = 100$, the graph for Sense Rent-a-Car will be above the graph for Herr's Rent-a-Car. Thus, Sense Rent-a-Car will be better if the miles driven exceeds 100. $(x > 100)$

47. Find the common ordered pairs in each table. The x-coordinates are the solutions.

49. Add or subtract to move variable terms to one side and constants to the other. Add like terms. Divide by the numerical coefficient of x.

51. Subtract c from both sides and then multiply both sides by $\dfrac{b}{a}$.

53. Subtraction is not the inverse of multiplication. Both sides should be divided by 4.

Exercises 3.5

1. $(5, +\infty)$

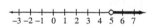

3. $(-\infty, -2]$

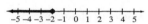

5. $[0, +\infty)$

7. $(-\infty, -1)$

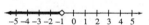

9. From the graph, $x > 2$.

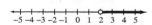

11. From the graph, $x < 1$.

13. From the graph, $x > 2$.

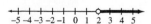

15. From the graph, $x < 4$.

17. $x > -2$

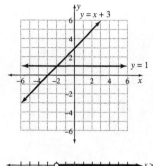

19. $x \geq 4$

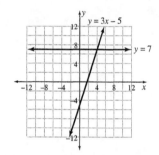

21. $x < 1$

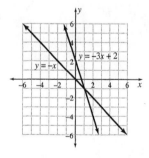

23. $x < 1$

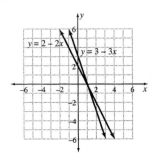

$\longleftarrow\!\!\!+\!\!+\!\!+\!\!+\!\!+\!\!\!\circ\!\!+\!\!+\!\!+\!\!+\!\!\!\longrightarrow\ x<1$
$-6\ \ -4\ \ -2\ \ \ 0\ \ \ 2\ \ \ 4\ \ \ 6$

25. $-3 > -3x + 3$

$-6 > -3x$

$2 < x$

$x > 2$

$\left(2, \infty\right)$

27. $0 < -3x + 3$

$3x < 3,\ \ x < 1,\ \ (-\infty, 1)$

29. $-x + 4 < \frac{1}{2}x + 1$

$3 < \frac{3}{2}x,\ \ 2 < x$

$x > 2,\ \ (2, +\infty)$

31. $-x + 4 > 0,\ \ 4 > x$

$x < 4,\ \ (-\infty, 4)$

33. $-1 < -3x + 2,\ \ 3x < 3$

$x < 1,\ \ (-\infty, 1)$

35. $-2 \le 2 - 2x,\ \ 2x \le 4$

$x \le 2,\ \ (-\infty, 2]$

37. $3 - 2x \le -5,\ \ 8 \le 2x,\ \ 4 \le x$

$x \ge 4,\ \ [4, +\infty)$

39. $-4 > x + 5,\ \ -9 > x$

$x < -9,\ \ (-\infty, -9)$

41. $1 - 4x < -4,\ \ 5 < 4x,\ \ \frac{5}{4} < x$

$x > 1\frac{1}{4},\ \ (1\frac{1}{4},\ +\infty)$

43. $-2 > 2x + 1,\ \ -3 > 2x,\ \ -\frac{3}{2} > x$

$x < -1\frac{1}{2},\ \ (-\infty,\ -1\frac{1}{2})$

45. $2x < x + 5,\ \ x < 5,\ \ (-\infty, 5)$

47. $4 - 2x \ge x - 2,\ \ 6 \ge 3x,\ \ 2 \ge x$

$x \le 2,\ \ (-\infty, 2]$

49. $3x - 4 < -2x + 1,\ \ 5x < 5,$

$x < 1,\ \ (-\infty, 1)$

51. $2(x + 3) < 5x,\ \ 2x + 6 < 5x$

$6 < 3x,\ \ 2 < x$

$x > 2,\ \ (2, +\infty)$

53. $\frac{1}{2}x > 4 + x,\ \ x > 8 + 2x$

$-x > 8,\ \ x < -8,\ \ (-\infty, -8)$

55. $-x > -\frac{1}{2}x + 1,\ \ -\frac{1}{2}x > 1$

$x < -2,\ \ (-\infty, -2)$

57. **a.** The point of intersection is $(16, 168)$.

    b. Let x be the number of people.

$9.25x + 20 \le 168$

    c. $9.25x + 20 \le 168$

$9.25x \le 148$

$x \le 16$

    Inequality: $0 < x \le 16$

    Interval: $(0, 16]$

    Graph:

$\longleftarrow\!\!+\!\!\!\circ\!\!+\!\!+\!\!+\!\!+\!\!+\!\!+\!\!+\!\!\bullet\!\!+\!\!\!\longrightarrow$
$2\ \ 0\ \ 2\ \ 4\ \ 6\ \ 8\ \ 10\ 12\ 14\ 16\ 18$

59. Let x be the number of children.

$\$5.50x + \$17 \le \$160$

$\$5.50x \le \$143,\ \ x \le 26$ children

61. Let x be the number of people.

$\$8.50x + \$350 \le \$1030$

$\$8.50x \le \$680,\ \ x \le 80$ people

63. Let x be the number of people.

$\$8.50x + \$350 < \$17.50x + \$100$

$\$8.50x < \$17.50x - \$250$

$-\$9.00x < -\$250$

$$x > \frac{250}{9} \approx 27.8$$

It will cost less to use Country Inn if there are more than 27 people attending the reception.

65. Let x be the final exam score.
$$\frac{78+84+72+5+x}{520} \geq 0.8$$
$$\frac{239+x}{520} \geq 0.8$$
$$239+x \geq 416$$
$$x \geq 177$$
Since there are only 150 points on the final, it is not possible for the student to earn a B.

67. Let x be the combined total of the three tests.
$$\frac{70+x+135}{520} \geq 0.9$$
$$\frac{x+205}{520} \geq 0.9$$
$$x+205 \geq 468$$
$$x \geq 263$$
The student must have a combined test total of at least 263 points.

69. Let x be the final exam score.
$$\frac{88+84+89+70+x}{100+100+100+70+200} \geq 0.90$$
$$\frac{331+x}{570} \geq 0.90, \quad 331+x \geq 513$$
$$x \geq 182 \text{ points}$$

71. If $a < b$, then $\dfrac{a}{c} > \dfrac{b}{c}$, where a and b are real numbers and $c < 0$.

73. The direction of the inequality reverses:
$+4 > +3$

Chapter 3 Review Exercises

1. Equivalent equations.

3. 2 variable equation and a conditional equation

5. Conditional equation

7. $4 = 5x - 11$

9. $\dfrac{x}{6} = 12$

11. $2x - 5 = x$

13. $2(7-x) = -4$

15. Three less than 4 times a number is 29.

17. The product of 2 and the difference between a number and 1, subtracted from 5, is 7 more than the number.

19. $-5(-1) + 4 = 9$?
$\quad\quad 5 + 4 = 9$?
$\quad\quad\quad\quad 9 = 9$ true

21. $8 - 7.5(0.4) = 5$?
$\quad 8 - 3 = 5$, true statement

23. $x + 3 = -4, \quad x = -7$

25. $3x = 27, \quad x = 9$

27. $\frac{1}{2}x = 12, \quad x = 24$

29. $4x - 2 = 22, \quad 4x = 24, \quad x = 6$

31. $-2x + 3 = 9, \quad -2x = 6, \quad x = -3$

33. $-2(x-4) = 18, \quad x - 4 = -9, \quad x = -5$

35. $5 - 2(x+1) = -11$
$\quad 5 - 2x - 2 = 3 - 2x = -11$
$\quad\quad -2x = -14, \quad x = 7$

37. $3x - 1 = x + 1, \quad 2x = 2, \quad x = 1$

39. $-2(x-3) = \dfrac{1}{2}x + 3$
$$-2x + 6 = \dfrac{1}{2}x + 3$$
$$-\dfrac{5}{2}x = -3$$
$$x = \left(-\dfrac{2}{5}\right)(-3)$$
$$x = \dfrac{6}{5} \text{ or } x = 1\dfrac{1}{5}$$

41. $12 - 1 = 11, \quad 12 + 1 = 13$

43. $15 + 3 = 18, \quad 15 + 6 = 21$

45. $9 = 10 - 1 = x - 1$
$\quad 11 = 10 + 1 = x + 1$

47. $-5 = -3 - 2 = x - 2$
$\quad -1 = -3 + 2 = x + 2$

49. $x + 2(x + 1) = 17$
$x + 2x + 2 = 17, \quad 3x + 2 = 17$
$3x = 15, \quad x = 5$

51. a From the table, when $y = 4$,
$x = 2$

 b. From the table, when $y = 13$,
$x = 5$

 c. From the table, when $y = 8$, x is between 3 and 4. Solving for x:
$8 = 3x - 2, \quad 10 = 3x, \quad x = \frac{10}{3}, \quad x = 3\frac{1}{3}$

 d. Extending the table, when $y = 16$,
$x = 6$.

53. a. From the graph, when $y = 0$,
$x = \{0, 4\}$

 b. From the graph, when $y = 6$, $x = \{1, 3\}$

 c. From the graph, when $y = 8$, $x = \{2\}$

 d. From the graph, when $y = 9$, $x = \{\ \}$
(the graph never reaches $y = 9$)

55. a. From the graph:
$3x - 3 = x + 1$ when $x = 2$

 b. From the graph:
$x + 1 = -2$ when $x = -3$

 c. From the graph:
$3x - 3 = -6$ when $x = -1$

57. Let y be distance and x be speed.
$y = 4x$

59. Let y be the tax and x be the cost of the meal.
$y = 0.075x$

61. Let y be the total cost and x be the number of credits.
$y = \$300x + \$150$

63. Let y be the remaining value and x be the number of visits.
$y = -\$10x + \$520$

65. $440 = 20(x + 3)$
$440 = 20x + 60$
$380 = 20x, \quad 19 = x$
19 seats in original row

67. $x + (x + 5) + (x + 10) = 60$
$3x + 15 = 60, \quad 3x = 45$
$x = 15, x + 5 = 20, x + 10 = 25$

69. a, b, c does not indicate that the numbers are 1 unit apart.

71. $2 < x - 3$, $5 < x$, Solution a, $x > 5$

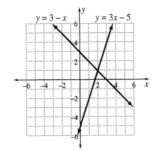

73. $15 - 2x < x - 6$
$21 < 3x$,
$7 < x$,
$x > 7$,
$(7, +\infty)$

75. $3x - 5 \leq 3 - x$,
$4x \leq 8$
$x \leq 2$,
$(-\infty, 2]$

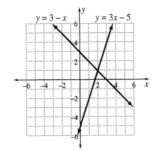

77. $1 < x - 4$,
$5 < x$
$x > 5$,
$(5, +\infty)$

79. $-\frac{1}{2}x > 8$,
$x < -16$,
$(-\infty, -16)$

81. $5 - 3x > 13$,
$-3x > 8$,
$x < -\frac{8}{3}$
$x < -2\frac{2}{3}$,
$(-\infty, -2\frac{2}{3})$

83. $13 \leq 7 - \frac{x}{3}$, $6 \leq -\frac{x}{3}$, $-18 \geq x$
$x \leq -18$, $(-\infty, -18]$

85. $-2x < 1 - x$, $-x < 1$,
$x > -1$, $(-1, +\infty)$

87. $x > 2x + 3, -3 > x,$
$x < -3, \ (-\infty, -3)$

89. $x + 1 \leq 3 - x, \ 2x \leq 2$
$x \leq 1, \ (-\infty, 1]$

91. $-2x + 2 > -2 - x, \ 4 > x,$
$x < 4, \ (-\infty, 4)$

93. $3x + 2 \geq 3 - 2x, \ 5x \geq 1$
$x \geq \frac{1}{5}, \ [\frac{1}{5}, +\infty)$

95. Student 1 has earned 244 of a possible 370 points so far.
$$\frac{244 + x}{370 + 150} \geq 0.80, \quad \frac{244 + x}{520} \geq 0.80$$
$244 + x \geq 416, \ x \geq 172$ points, not possible.

97. $\$25x + \$550 \leq \$3500$
$\$25x \leq \$2950, \ x \leq 118$ people

99. $\$12x + \$250 < \$8x + \$320$
$\$4x < \$70, \ x < 17.5$ feet

Chapter 3 Test

1. Let x be a number:
$\frac{1}{2}x + 6 = 15, \ \frac{1}{2}x = 9, \ x = 18$

2. Let x be a number:
$3x - 7 = -31, \ 3x = -24, \ x = -8$

3. Let x be input, y be output:
$y = \frac{1}{3}x$

4. Let x be input, y be output:
$y = 2x - 2$

5. Let x be the first integer:
$x + (x + 1) + (x + 2) + (x + 3) = -74$
$4x + 6 = -74, \ 4x = -80, \ x = -20$

6. $x + 8 = -3, \ x = -11$

7. $4 - x = 5, \ -x = 1, \ x = -1$

8. $\frac{2}{5}x = 30, \ x = \frac{30 \cdot 5}{2}, \ x = 75$

9. $-6x = 3, \ x = \frac{3}{-6}, \ x = -\frac{1}{2}$

10. $5 - 2x = 3, \ 2 = 2x, \ x = 1$

11. $\frac{1}{2}x + 5 = -3, \ \frac{1}{2}x = -8, \ x = -16$

12. $2(x - 2) = -x - 1, \ 2x - 4 = -x - 1$
$3x = 3, \ x = 1$

13. $4(x - 3) = 6, \ 4x - 12 = 6$
$4x = 18, \ x = 4\frac{1}{2}$

14. $4 - 2(x - 4) = 2x, \ 4 - 2x + 8 = 2x$
$12 - 2x = 2x, \ 12 = 4x, \ x = 3$

15. $-2(x - 3) = -0.5(x - 6)$
$-2x + 6 = -0.5x + 3, \ 3 = 1.5x$
$x = 2$

16. solution $(1, 3)$

17. Zero is between -1 and 1 so x is between 2 and 3. The line is straight so because zero is half-way between -1 and 1 the value of x will be half-way between 2 and 3.

18. Find the point where the graph crosses the x-axis.

19. a. $(1, -2)$

     b. $-2 = 2(1) - 4$?
        $-2 = 2 - 4$, checks
        $-2 = -(1) - 1$?
        $-2 = -1 - 1$, checks

     c. $2x - 4 = -x - 1, \ 3x = 3, x = 1$

     d. It is the x value of the point of intersection.

     e. $2x - 4 > -x - 1$
        $3x - 4 > -1$
          $3x > 3$
            $x > 1$

20. $2x + 5 < -3, \ 2x < -8, \ x < -4, \ (-\infty, -4)$

21. $3 - 2x > 11, \ -8 > 2x, \ -4 > x$
$x < -4, \ (-\infty, -4)$

22. $2x + 8 \geq \frac{1}{2}(x + 1)$
$4x + 16 \geq x + 1, \ 3x \geq -15$
$x \geq -5, \ [-5, +\infty)$

23. a. Let x be the number of therms and y be the
        total cost in dollars.

x	$y = 0.615x + 4$
0	$0.615(0) + 4 = 4$
20	$0.615(20) + 4 = 16.30$
40	$0.615(40) + 4 = 28.60$
60	$0.615(60) + 4 = 40.90$
80	$0.615(80) + 4 = 53.20$
100	$0.615(100) + 4 = 65.50$

b.

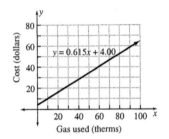

c. $y = \$0.615x + \$4$

d. $\$47.05 = \$0.615x + \$4$
        $\$43.05 = \$0.615x$
        $x = 70$ therms

Chapter 4

1. Interest in terms of principal, rate and time.

3. Rate in terms of distance and time.

5. Grade percent in terms of tests 1, 2, & 3, homework, final exam, and total points possible.

7. $A = lw$

9. $A = \pi r^2$

11. $A = \frac{1}{2}bh$

13. $P = 2l + 2w$

15. $C = fm$

17. $p = 5n$,

$$\frac{p}{5} = \frac{5n}{5}$$

$$n = \frac{p}{5}$$

19. $A - P = H$,
$A - P + P = H + P$,
$A = H + P$

21. $C = 2\pi r$,

$$\frac{C}{2\pi} = \frac{2\pi r}{2\pi},$$

$$r = \frac{C}{2\pi}$$

23. $A = bh$,

$$\frac{A}{b} = \frac{bh}{b},$$

$$h = \frac{A}{b}$$

25. $I = prt$,

$$\frac{I}{pr} = \frac{prt}{pr},$$

$$t = \frac{I}{pr}$$

27. $C = \pi d$,

$$\frac{C}{\pi} = \frac{\pi d}{\pi},$$

$$d = \frac{C}{\pi}$$

29. $P = R - C$,
$P + C = R$,
$P + C - P = R - P$,
$C = R - P$

31. $PV = nRT$,

$$\frac{PV}{RT} = \frac{nRT}{RT},$$

$$n = \frac{PV}{RT}$$

33. $C_1 V_1 = C_2 V_2$,

$$\frac{C_1 V_1}{C_1} = \frac{C_2 V_2}{C_1},$$

$$V_1 = \frac{C_2 V_2}{C_1}$$

35. $P = a + b + c$,
$P - a - b = a + b + c - a - b$,
$c = P - a - b$

37. $A = \frac{1}{2}h(a + b)$,

$2A = 2 \cdot \frac{1}{2}h(a + b)$

$2A = h(a + b)$

$$\frac{2A}{a + b} = \frac{h(a + b)}{a + b}$$

$$h = \frac{2A}{a + b}$$

39. $V = \frac{1}{3}\pi r^2 h$,

$3V = 3 \cdot \frac{1}{3}\pi r^2 h$,

$3V = \pi r^2 h$,

$$\frac{3V}{\pi h} = \frac{\pi r^2 h}{\pi h}$$

$$r^2 = \frac{3V}{\pi h}$$

41. $x = \dfrac{-b}{2a}$,

$2ax = 2a\dfrac{-b}{2a}$,

$2ax = -b$,

$(-1)2ax = (-1)(-b)$

$b = -2ax$

43. $y = mx + b$,

$y - mx = mx + b - mx$

$b = y - mx$

45. $d^2 = \dfrac{3h}{2}$,

$\dfrac{2d^2}{3} = \dfrac{2}{3} \cdot \dfrac{3h}{2}$,

$h = \dfrac{2d^2}{3}$

47. $t^2 = \dfrac{2d}{g}$,

$gt^2 = g \cdot \dfrac{2d}{g}$,

$gt^2 = 2d$,

$\dfrac{gt^2}{t^2} = \dfrac{2d}{t^2}$,

$g = \dfrac{2d}{t^2}$

49. a. $A = P + Prt$,

$A - P = Prt$

$r = \dfrac{A\text{-}P}{Pt}$

b. $A = \$11{,}050$, $P = \$10{,}000$

$t = 2$ years

$r = \dfrac{11{,}050\text{-}10{,}000}{10{,}000 \cdot 2} = \dfrac{1{,}050}{20{,}000}$

$r = 0.0525$ or $5\frac{1}{4}\%$

51. a. $C = \frac{5}{9}(F\text{-}32)$,

$\frac{9}{5}C = F\text{-}32$

$F = \frac{9}{5}C + 32$

b. $C = 37°$,

$F = \frac{9}{5} \cdot 37 + 32$, $F = 98.6°$

53. a. $H = 0.8(200 - 140)$,

$H = 0.8(60)$,

$H = 48$

b. $H = 0.8(200 - A)$

$200 - A = \dfrac{H}{0.8}$,

$A = 200\text{-}\dfrac{H}{0.8}$

c. $H = 24$,

$A = 200\text{-}\dfrac{24}{0.8}$,

$A = 170$

55. a. $y = mx + b$,

$b = y - mx$

b. $b = 4 - 2(3)$, $b = -2$

c. $b = 4 - (-2)(3)$, $b = 10$

d. $b = 4 - \frac{1}{2}(3)$, $b = 2\frac{1}{2}$

e. $b = 4 - (-\frac{1}{2})(3)$, $b = 5\frac{1}{2}$

57. $xy = -6$,

$\dfrac{xy}{x} = \dfrac{-6}{x}$,

$y = \dfrac{-6}{x}$

59. $2x - y = 3$

$2x - y - 2x = 3 - 2x$

$-y = 3 - 2x$

$(-1)(-y) = (-1)(3 - 2x)$

$y = -3 + 2x$

$y = 2x - 3$

61. $2x - 3y - 4 = 0$

$2x - 3y - 4 + 4 = 0 + 4$

$2x - 3y = 4$

$2x - 3y - 2x = 4 - 2x$

$-3y = 4 - 2x$

$\dfrac{-3y}{-3} = \dfrac{4 - 2x}{-3}$

$y = \dfrac{2x - 4}{3}$

$y = \dfrac{2}{3}x - \dfrac{4}{3}$

63. $2x - 3y - 4 = 0$

$2x + 3y - 2x = 9 - 2x$

$3y = 9 - 2x$

$\dfrac{3y}{3} = \dfrac{9 - 2x}{3}$

$y = \dfrac{-2x + 9}{3}$

$y = -\dfrac{2}{3}x + 3$

Exercises 4.2

1. a. There is exactly one output for each input, it is a function; inputs are the set of first numbers, {-6, -5, 5, 6}; outputs are the set of second numbers {5, 6}.

    b. Not a function; the input 5 has two different outputs, the input 6 has two different outputs.

3. a. Function, inputs {2, 3, 4} output $\{\frac{1}{2}, \frac{1}{3}, \frac{1}{4}\}$

    b. Not a function

5. Function, input {Eden, Tuckman, McClintock}, output {Barbara}

7. Not a function

9. (a) For each input, there is exactly one output.

11. Long distance is billed at cost per unit of time. input {units of time}, output {cost of call}

13. The tilt of the Earth gives long December days in the southern hemisphere and short days in the northern. input {distance from equator}, output {hours of sunlight}

15. Let x be the number of hours worked, and A be the amount earned (in dollars).

$A(x) = 8x$

17. C = circumference, d = diameter,

$C(d) = \pi d$

19.

x	$f(x) = 2x - 1$
-2	2(-2) - 1 = -5
-1	2(-1) - 1 = -3
0	2(0) - 1 = -1
1	2(1) - 1 = 1
2	2(2) - 1 = 3

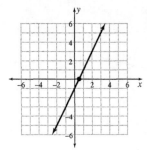

21.

x	$f(x) = 2 - 3x$
-2	2 - 3(-2) = 8
-1	2 - 3(-1) = 5
0	2 - 3(0) = 2
1	2 - 3(1) = -1
2	2 - 3(2) = -4

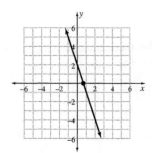

23.

x	$f(x) = \frac{1}{4}x + 1$
-2	$\frac{1}{4}(-2) + 1 = \frac{1}{2}$
-1	$\frac{1}{4}(-1) + 1 = \frac{3}{4}$
0	$\frac{1}{4}(0) + 1 = 1$
1	$\frac{1}{4}(1) + 1 = 1\frac{1}{4}$
2	$\frac{1}{4}(2) + 1 = 1\frac{1}{2}$

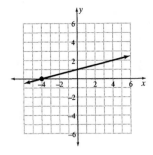

25. $h(4) = \left|\frac{1}{2}(4)\right| = |2| = 2$

$h(-4) = \left|\frac{1}{2}(-4)\right| = |-2| = 2$

27. $H(4) = 4 - 4^2 = 4 - 16 = -12$
$H(-4) = (-4) - (-4)^2 = -4 - 16 = -20$

29. $g(-2) = (-2)^2 + 1 = 4 + 1 = 5$
$g(1) = (1)^2 + 1 = 1 + 1 = 2$

31. $G(-2) = 2 - (-2)^2 = 2 - 4 = -2$
$G(1) = 2 - (1)^2 = 2 - 1 = 1$

33. Not a function, there are 2 values of y for some x's. Fails the vertical line test.

35. Function

37. Not a function, there are 2 values of y for some x's. Fails the vertical line test.

39. Function of x. One value of y for each value of x.

41. Function of x. One value of y for each value of x.

43. Function of x. One value of y for each value of x.

45. Not a function of x. Two possible values of y for each x.

47. Not a function of x. Infinite number of values of y for each x.

49. Substitute a for every x in the equation defined by $f(x)$.

51. Answers will vary, multiplication is generally written $f \cdot x$ or fx

53. a. It is the input.

   b. The values for both variables are positive in the first quadrant. Negative gallons and negative cost make no sense.

   c. We connect the points in the graph because fractional or decimal parts of 1000 gallons are possible.

55. a. $f(0) = 0$

   b. $f(20) = \$5$

   c. $f(100) = \$5$

   d. $f(200) = 0.03(200) = \$6$

   e. $f(350) = 0.03(350) = \$10.50$

   f. $f(1000) = 0.03(1000) = \$30$

   g. $f(1100) = 0.03(1100) = \$33$

   h. $f(1500) = \$40$

   i. $f(x)$ is 3% of the input.

   j. $x = 0$

   k. Cannot tell; many inputs have this output value.

   l. $0.03x = 12$
   $$x = \frac{12}{0.03}$$
   $x = 400$
   The input is \$400.

   m. $0.03x = 21$
   $$x = \frac{21}{0.03}$$
   $x = 700$
   The input is \$700.

n. $0.03x = 27$

$$x = \frac{27}{0.03}$$

$$x = 900$$

The input is $900.

o. $0.03x = 36$

$$x = \frac{36}{0.03}$$

$$x = \$1200$$

p.   Cannot tell; many inputs have this output value.

57. Let I be the interest earned, P be the principal, and i be the interest rate.

$$I(P,i) = P \cdot i$$

59. Let A be the area, b be the length of the base, and h be the height.

$$A(b,h) = b \cdot h$$

Exercises 4.3

1. a.   Line rises from left to right, positive slope. Rise = 3, run = 1, slope = $\frac{3}{1} = 3$

   b.   Line rises from left to right, positive slope. Rise = 2, run = 3, slope = $\frac{2}{3}$

   c.   Horizontal line, zero slope.   Rise = 0, run = 3,   slope = 0

3. a.   Line descends from left to right, negative slope. Rise = -2, run = 1, slope = $\frac{-2}{1} = -2$

   b.   Line rises from left to right, positive slope. Rise = 3, run = 2, slope = $\frac{3}{2}$

   c.   Vertical line, undefined slope. Rise = 4, run = 0, slope = $\frac{4}{0}$, undefined

5. a.   Line descends from left to right, negative slope. Rise = -50, run = 1, slope = $\frac{-50}{1} = -50$

   b.   Line rises from left to right, positive slope. Rise = 40, run = 1, slope = $\frac{40}{1} = 40$

   c.   Line descends from left to right, negative slope. Rise = -20, run = 1, slope = $\frac{-20}{1} = -20$

7.   Slope = $\dfrac{3-2}{4-0} = \dfrac{1}{4}$

9.   Slope = $\dfrac{-4-3}{0-(-2)} = \dfrac{-7}{2} = -3.5$

11.   Slope = $\dfrac{4-3}{4-4} = \dfrac{1}{0}$,   undefined

13.   Slope = $\dfrac{2-2}{-2-0} = \dfrac{0}{-2} = 0$

15.   Slope = $\dfrac{-1-3}{4-(-2)} = \dfrac{-4}{6} = \dfrac{-2}{3}$

17.   Slope = $\dfrac{-3-(-3)}{4-2} = \dfrac{0}{2} = 0$

19.   Slope = $\dfrac{0-4}{5-0} = \dfrac{-4}{5}$

21.   Slope = $\dfrac{4-(-4)}{3-3} = \dfrac{8}{0}$,   *undefined*

23. As x values increase y values decrease, negative slope. Δx between each input is 1, Δy between each output is -3, linear function. Slope = -3, no units are given.

25. As x values increase y values increase, positive slope. Δx between each input is 1, Δy between each output is 7, linear function. Slope = 7, no units are given.

27. As x values increase y values increase, positive slope. Δx between each input is 2, Δy between each output is 18, linear function. Slope = $\frac{18}{2} = 9$.

Units are $\dfrac{\text{earnings}}{\text{hour}}$.

29. As x values increase y values increase, positive slope. Δx between each input is 1, Δy between each output is $0.50, linear function. Slope = $0.50.

Units are $\dfrac{\text{cost}}{\text{kilogram}}$.

31. As x values increase y values increase, positive slope. Δx between each input is 2, Δy between each output is $0.64, linear function.

    Slope $\dfrac{\$0.64}{2} = \$0.32$. Units are $\dfrac{\text{cost}}{\text{pound}}$.

33. As x values increase y values increase, positive slope. Δx between each input is 1, Δy between each output varies, non-linear function. Units are $\dfrac{\text{ft}}{\text{sec}}$.

35. As x values increase y values decrease, negative slope. Δx between each input and Δy between each output appear to vary, a comparison of slope between each point reveals a linear function. Slope

    $= \dfrac{-\$2.50}{10} = \dfrac{-\$1.25}{5} = -\$0.25$. Units are $\dfrac{\$\text{ value}}{\text{copy}}$.

37. As x values increase y values increase, positive slope. Δx between each input is 1, Δy between each output varies, non-linear function. Units are $\dfrac{\text{miles}}{\text{hour}}$.

39. $\dfrac{y_2 - y_1}{x_2 - x_1} = \dfrac{d - b}{c - a}$

41. $\dfrac{y_2 - y_1}{x_2 - x_1} = \dfrac{b - 0}{0 - a} = -\dfrac{b}{a}$

43. $\dfrac{y_2 - y_1}{x_2 - x_1} = \dfrac{c - b}{a - a} = \dfrac{c - b}{0}$, *undefined*

45. Rise = 5 ft, run = 10 ft

    Slope $= \dfrac{5 \text{ ft}}{10 \text{ ft}} = \dfrac{1}{2}$

47. Rise = 7 ft, run $= \dfrac{28 \text{ ft}}{2} = 14 \text{ ft}$

    Slope $= \dfrac{7 \text{ ft}}{14 \text{ ft}} = \dfrac{1}{2}$

49. a.

gallons (g)	cost (c)
0	0
1	$1.55
2	2($1.55) = $3.10

    b. Slope = $1.55 $\dfrac{\text{cost}}{\text{gallon}}$

    c. c = $1.55g

51. a.

hours (h)	earnings (e)
0	0
1	$6.25
2	2($6.25) = $12.50

    b. Slope = $6.25 $\dfrac{\$ \text{ earnings}}{\text{hour}}$

    c. e = $6.25h

53. a.

hours (x)	distance (d) in kilometers
0	0
1	80
2	2(80) = 160

    b. Slope = 80 $\dfrac{\text{km}}{\text{hr}}$

    c. d = 80x

55. Slope tells you relative direction up or down as you move from left to right and how steep the graph is as it climbs or descends.

57. Slope will be negative when the y values get smaller as the x values get larger.

59. Select two points on the line. Going left to right on the graph, count the change in y and divide by the change in x. Write the appropriate sign.

61. Write the slope in the form $\dfrac{a}{b}$ where $b > 0$.

From a point, go a units up if a is positive and down if a is negative and go b units horizontally to the right to find a second point. Draw a line between the two points.

63. $-\dfrac{1}{3}, -1, -\dfrac{3}{1}$

65. $\dfrac{2}{5}, \dfrac{1}{2}, \dfrac{6}{5}$

67. $-\dfrac{1}{2}, -\dfrac{2}{3}, -\dfrac{3}{2}, -2$

69. The answer shows the change in x over the change in y. Slope is the reciprocal; change in y over the change in x.

71. **a.**

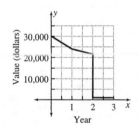

b. $\dfrac{30,000 - 24,000}{0 - 1} = -\$6,000$

c. $\dfrac{24,000 - 21,600}{1 - 2} = -\$2,400$

d. Extreme loss of value happened suddenly.

e. $\dfrac{1000 - 1000}{2 - 3} = 0$

f. Answers will vary. One possible answer: After driving it for 2 years, she had a wreck in the third year.

Mid-Chapter 4 Test

1. Temperature in Celsius in terms of temperature in Fahrenheit.

2. Volume of a sphere in terms of radius.

3. $3 = 4(-2) + b$,
$3 = -8 + b$,
$b = 11$

4. $-4 = \frac{2}{3}(-6) + b$,
$-4 = -4 + b$,
$b = 0$

5. $C = K - 273$,
$K = C + 273$

6. $d^2 = \dfrac{3h}{8}$,

$\dfrac{8}{3} \cdot d^2 = \dfrac{8}{3} \cdot \dfrac{3h}{8}$,

$h = \dfrac{8d^2}{3}$

7. $l = a + (n - 1)d$,
$l - a = (n - 1)d$,
$d = \dfrac{l\text{-}a}{n\text{-}1}$

8. **a.** $f(6) = 3(6) = 18$

     b. $f(6) = \frac{1}{2}(6) + 2 = 3 + 2 = 5$

     c. $f(6) = \frac{1}{2}(6 + 2) = \frac{1}{2}(8) = 4$

     d. $f(6) = (6)^2 - 2 = 36 - 2 = 34$

9.

10. **a.** Yes, one value of y for each x.

     b. $\{2, 3, 4, 5, 6, 7\}$

     c. $\{3, 4\}$

11. a.

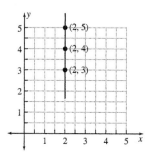

b. exactly one

c. one input, three outputs

d. Does not pass.

12. $m = \dfrac{y_2 - y_1}{x^2 - x_1} = \dfrac{6-8}{4-(-3)} = \dfrac{-2}{7} = -\dfrac{2}{7}$

13. $m = \dfrac{y_2 - y_1}{x_2 - x_1} = \dfrac{8-8}{4-(-3)} = \dfrac{0}{7} = 0$

14. $m = \dfrac{y_2 - y_1}{x_2 - x_1} = \dfrac{21-25}{3-1} = \dfrac{-4}{2} = \dfrac{-2}{1} = -2$

This indicates that the number of survivors decreases by 2 each episode.

15. BA:

$m = \dfrac{4-0}{-4-0} = \dfrac{4}{-4} = -1$

BC:

$m = \dfrac{4-0}{4-0} = \dfrac{4}{4} = 1$

Exercises 4.4

1. x-intercept:
$3x + 5y = 15$
$3x + 5(0) = 15$
$3x = 15$
$x = 5$
The x-intercept point is $(5,0)$.
y-intercept:
$3x + 5y = 15$
$3(0) + 5y = 15$
$5y = 15$
$y = 3$
The y-intercept point is $(0,3)$.

3. x-intercept:
$y = 3x - 24$
$0 = 3x - 24$
$24 = 3x$
$8 = x$
The x-intercept point is $(8,0)$.
y-intercept:
$y = 3x - 24$
$y = 3(0) - 24$
$y = -24$
The y-intercept point is $(0,-24)$.

5. x-intercept:
$x = 6 - 2y$
$x = 6 - 2(0)$
$x = 6$
The x-intercept point is $(6,0)$.
y-intercept:
$x = 6 - 2y$
$0 = 6 - 2y$
$2y = 6$
$y = 3$
The y-intercept point is $(0,3)$.

7. x-intercept :
$4x - 3y = 12$
$4x - 3(0) = 12$
$4x = 12$
$x = 3$
The x-intercept point is $(3,0)$.
y-intercept:
$4x - 3y = 12$
$4(0) - 3y = 12$
$-3y = 12$
$y = -4$
The y-intercept point is $(0,-4)$.

9. $f(x) = 0$, because $y = 0$ and $y = f(x)$.

11. $f(0) = 2(0) - 1 = -1$
$f(x) = 0$
$2x - 1 = 0$
$2x = 1$
$x = \dfrac{1}{2}$

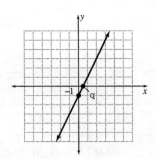

13. $f(0) = 2 - 3(0) = 2$

$$f(x) = 0$$

$$2 - 3x = 0$$

$$2 = 3x$$

$$\frac{2}{3} = x \text{ or } x = \frac{2}{3}$$

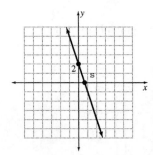

15. $f(0) = \frac{1}{4}(0) + 1 = 1$

$$f(x) = 0$$

$$\frac{1}{4}x + 1 = 0$$

$$\frac{1}{4}x = -1$$

$$x = -4$$

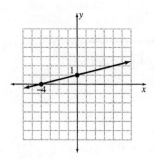

17. $f(x) = 0$
$3(x - 4) + x = 0,$
$3x - 12 + x = 0,$
$4x - 12 = 0$
$4x = 12,$
$x = 3$

19. $f(x) = 0,$
$9 - 3(x - 1) = 0,$
$9 - 3x + 3 = 0,$
$12 - 3x = 0,$
$12 = 3x$
$x = 4$

21. $f(x) = 0,$
$2x + 3(x - 5) = 0,$
$2x + 3x - 15 = 0,$
$5x - 15 = 0,$
$5x = 15,$
$x = 3$

23. $f(x) = 0,$
$115x + 48 = 0$
$115x = -48,$
$$x = \frac{-48}{115}$$
This value has no meaning.
$f(0) = 115(0) + 48 = 48$
This value is the fees included in the tuition.

25. $f(x) = 0,$
$19.50 - 0.75x = 0,$
$19.50 = 0.75x,$
$x = 26$
This is the total number of minutes the phone card will buy.
$f(0) = 19.50 - 0.75(0) = 19.50$
This is the original value of the phone card.

27. a. $y = \frac{1}{2}x$, $y = 2x$, $y = 4x$

    b. $y = -\frac{1}{3}x$, $y = -x$, $y = -3x$

    c. $y = x$, $y = 1.5x$, $y = 2.5x$, $y = 3x$

    d. $y = -0.25x$, $y = -0.5x$, $y = -x$,
        $y = -4x$

29. $m = 2$, $b = -\frac{1}{2}$

31. $m = -4$, $b = 15$

33. $m = -\frac{3}{4}$, $b = 0$

35. $2x = y + 4,$
    $y = 2x - 4,$
    $m = 2$, $b = -4$

37. $2x + 3y = 12$,
  $3y = -2x + 12$,
  $y = -\frac{2}{3}x + 4$
  $m = -\frac{2}{3}, \quad b = 4$

39. $5y - 2x = 10$,
  $5y = 2x + 10$,
  $y = \frac{2}{5}x + 2$
  $m = \frac{2}{5}, \quad b = 2$

41. $x - 4y = 4$,
  $4y = x - 4$,
  $y = \frac{1}{4}x - 1$
  $m = \frac{1}{4}, \quad b = -1$

43. $m = -0.30, \ b = 12$

45. $m = 55, \ b = 0$

47. $m = 2\pi, \ b = 8$

49. $m = 2.98, \ b = 0.50$

51. $m = -0.29, \ b = 50$

53. $H = 0.8(200 - A) = 160 - 0.8A$
  $m = -0.8, \ b = 160$

55. $C = 65 + 0.15(d - 100)$
  $C = 65 + 0.15d - 15 = 0.15d + 50$
  $m = 0.15, \ b = 50$

57. a.

Hours (x)	Cost (c)
0	$3
1	$1(1) + $3 = $4
2	$1(2) + $3 = $5

b. Slope = cost per hour = $1

c. When $x = 0$, $c = \$3$. This is the initial fee.

d. c = $1x + $3

59. a.

Meal Cost (x)	Total Cost (c)
0	0
$1	$1 + 0.15($1) = $1.15
$2	$2 + 0.15($2) = $2.30

b. Slope $= 1.15 = \dfrac{\text{total cost}}{\text{meal cost}}$

c. When $x = 0$, $c = 0$. This is the tip for a free meal.

d. c = 1.15x

NOTE: The following graph is for #61-67 (odd)

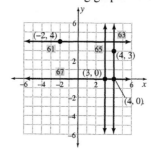

61. (See graph above)
  $m = 0 \,(horizontal\ line), \ b = 4$
  $y = 4$

63. (See graph above)
  m is undefined $(vertical\ line)$,
  no y-intercept
  $x = 4$

65. (See graph above)
  m is undefined $(vertical\ line)$,
  no y-intercept
  $x = 3$

67. (See graph above)
  $m = 0, \ b = 0, \ y = 0$ (x-axis)

NOTE: The following graph is for #69 and #71

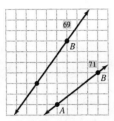

NOTE: The following graph is for #73-77 (odd)

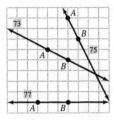

NOTE: The following graph is for #79 and #81

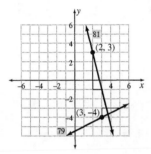

NOTE: The following graph is for #83-87 (odd)

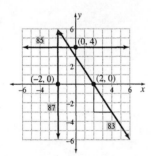

89. Let $y = 0$, or $f(x) = 0$, and solve for x.

91. a. x = the x-coordinate of the ordered pair.

   b. x = the x-intercept.

93. a. $10.50 = -0.15x + 15.00$
         $-4.50 = -0.15x$
           $30 = x$ or $x = 30$

   b. $y = -0.15(0) + 15.00$
         $y = 15.00$

c. $0 = -0.15x + 15.00$
      $0.15x = 15.00$
          $x = 100$

d. $6.75 = -0.15x + 15$
      $-8.25 = -0.15x$
         $55 = x$
      There have been 55 copies made.

e. The equation in part (c) finds the x-intercept.

f. The equation in part (b) finds the y-intercept.

95. a.

x, min	y, $
0	12
10	12 - 0.30(10) = 9
20	12 - 0.30(20) = 6
30	12 - 0.30(30) = 3
40	12 - 0.30(40) = 0
50	Not possible

b. $3.00

c. 10 minutes

d. $2 = 12 - 0.30x$
   $-10 = -0.30x$
   $\dfrac{-10}{-0.30} = x$, $x = 33\frac{1}{3}$ minutes

e. 40 minutes

f. Represents the total minutes available on the card.

g. $12

h. The initial cost of the card.

i. $y = -\$0.30x + \$12$

Exercises 4.5

1. $y = \frac{1}{2}x + 3$

3. $y = \frac{2}{3}x - 2$

5. $y = 5x + \frac{1}{4}$

7. $y = -\frac{3}{2}x + 1$

9. -1 = 4(3) + b,
    -1 - 12 = b,
    b = -13
    $y = 4x - 13$

11. -2 = (-1)(4) + b,
    -2 + 4 = b,
    b = 2
    $y = -x + 2$

13. $4 = (\frac{1}{2})(2) + b,$
    $4 - 1 = b,$
    $b = 3$
    $y = \frac{1}{2}x + 3$

15. $3 = (\frac{4}{5})(-10) + b,$
    $3 + 8 = b,$
    $b = 11$
    $y = \frac{4}{5}x + 11$

17. $1 = \frac{5}{3}(-3) + b$
    $1 = -5 + b$
    $1 + 5 = b$
    $6 = b$
    $y = \frac{5}{3}x + 6$

19. 3 = (-2)(1.5) + b,
    3 + 3 = b,
    b = 6
    $y = -2x + 6$

21. $m = \frac{9-1}{3-1} = \frac{8}{2} = 4$
    1 = (4)(1) + b,
    1 - 4 = b,
    b = -3
    $y = 4x - 3$

23. $m = \frac{-8-(-2)}{5-2} = \frac{-6}{3} = -2$
    -2 = (-2)(2) + b,
    -2 + 4 = b,
    b = 2
    $y = -2x + 2$

25. $m = \frac{-1-1}{0-(-3)} = \frac{-2}{3}$
    b = -1, from (0, -1)
    $y = -\frac{2}{3}x - 1$

27. $m = \frac{0-6}{10-13} = \frac{-6}{-3} = 2$
    6 = 2(13) + b,
    6 - 26 = b,
    b = -20
    $y = 2x - 20$

29. $m = \frac{-2-6}{-4-(-5)} = \frac{-8}{1} = -8$
    6 = (-8)(-5) + b,
    6 - 40 = b,
    b = -34
    $y = -8x - 34$

31. $m = \frac{3-2}{3-5} = \frac{1}{-2}$
    $2 = -\frac{1}{2}(5) + b,$
    $2 + \frac{5}{2} = b,$
    $b = 4\frac{1}{2}$
    $y = -\frac{1}{2}x + 4\frac{1}{2}$

33. Slope = -1, y-intercept = 1
    y = -x + 1

35. Slope = 2, y-intercept = 2
    y = 2x + 2

37. Slope = 0, y-intercept = -1
    y = -1

39. Slope = $\frac{1}{30}$, C-intercept = 1
    C = $\frac{1}{30}$n + 1

41. Slope = $\frac{-200}{50} = -4$,
    t - intercept = 40

$$0 = -4(40) + b$$
$$0 = -160 + b$$
$$160 = b$$
$$C = -4t + 160$$

43. *GH*

45. *NO*

47. *HI*:
$$m = \frac{0 - (-1)}{-3 - (-6)} = \frac{1}{3}$$

$$0 = \frac{1}{3}(-3) + b$$

$$1 = b$$

$$y = \frac{1}{3}x + 1$$

JK:
$$m = \frac{-3 - (-2)}{-5 - (-2)} = \frac{-1}{-3} = \frac{1}{3}$$

$$-2 = \frac{1}{3}(-2) + b$$

$$-2 = -\frac{2}{3} + b$$

$$-\frac{4}{3} = b$$

$$y = \frac{1}{3}x - \frac{4}{3}$$

LM:
$m = $ undefined (vertical line)
$x = -1$

NO:
$m = $ undefined (vertical line)
$x = 1$

49. **a., b., c.** are in the correct form.

d. $\quad 2y = x + 4, \quad y = \frac{1}{2}x + 2$

a and c have slopes that multiply to -1 so they are perpendicular.

b and d have the same slope so they are parallel.

51. **a.** is in the correct form.

b. $\quad x - \frac{1}{3}y = 6,$
$$x - 6 = \frac{1}{3}y,$$
$$y = 3x - 18$$

c. $\quad 3y - x = 2,$
$$3y = x + 2,$$
$$y = \frac{1}{3}x + \frac{2}{3}$$

d. $\quad y + \frac{1}{3}x = 4,$
$$y = -\frac{1}{3}x + 4$$

a and c have the same slope so they are parallel.

b and d have slopes that multiply to -1 so they are perpendicular.

53. $b = 0, \ y = 4x$

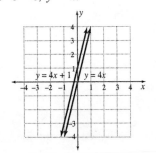

55. $b = 0, \ 2 \cdot m = -1, \ m = -\frac{1}{2}$

$$y = -\frac{1}{2}x$$

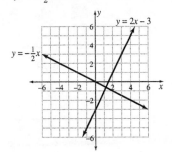

57. $b = 4, \ m = \frac{1}{3}$

$$y = \frac{1}{3}x + 4$$

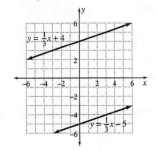

59. $b = -2$, $(-\frac{3}{4})m = -1$, $m = \frac{4}{3}$

$y = \frac{4}{3}x - 2$

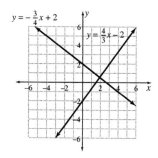

61. Find the slope from $m = \dfrac{y_2 - y_1}{x_2 - x_1}$, then substitute one ordered pair and m into $b = y - mx$ to find b. Place m and b into the equation $y = mx + b$.

63. $c =$ slope, $d =$ y-intercept.

65. Slope $= -\dfrac{b}{a}$

67. a. $m = \dfrac{b-0}{0-a} = -\dfrac{b}{a}$

   b. The opposite of the y-intercept divided by the x-intercept.

   c. $m = -\dfrac{-4}{-5} = -\dfrac{4}{5}$

       $m = \dfrac{-4-0}{0-(-5)} = \dfrac{-4}{5} = -\dfrac{4}{5}$

69. Data points are (7, \$6.01), (8, \$6.79)

$m = \dfrac{y_2 - y_1}{x_2 - x_1} = \dfrac{6.79 - 6.01}{8 - 7} = \$0.78$

$b = y - mx$,
$b = \$6.01 - \$0.78(7)$,
$b = \$0.55$
$y = \$0.78x + \$0.55$

71. a. Freezing ordered pair (0, 32)
       Boiling ordered pair (100, 212)

       Slope $= \dfrac{212 - 32}{100 - 0} = \dfrac{180}{100} = \dfrac{9}{5} \dfrac{°F}{°C}$

   b. The y-intercept is $b = 32$.

   c. $F = \dfrac{9}{5}C + 32$

73. a. Rental per hour, \$2

   b. Prepaid amount, \$50

   c. $y = -\$2x + \$50$

75. a. Rental per hour, \$42

   b. Insurance fee, \$28

   c. $y = \$42x + \$28$

77. per hour, percent of

79. Slope did not change; parallel line.
C = \$0.25n + \$45

81. Slope increased; steeper line.
C = \$1.60g

83. Slope did not change; parallel line.
C = 0.01x + \$5

Exercises 4.6

1. Multiply both sides by -1 and reverse the inequality sign.

3. Multiply both sides by -2 and reverse the inequality sign.

5. Subtract 2x from both sides, divide both sides by -3 and reverse the inequality sign.

7. -2 + 3 = 1, 1 < 3, (-2, 3) is not a solution

   4 + 0 = 4, 4 > 3, (4, 0) is a solution

   1 + 4 = 5, 5 > 3, (1, 4) is a solution

9. $\frac{1}{2}(-2) + 3 = 2$, $2 \geq 2$, (-2, 3) is a solution.

   $\frac{1}{2}(4) + (-2) = 0$, $0 < 2$, (4, -2) is not a solution

   $\frac{1}{2}(6) + (-1) = 2$, $2 \geq 2$, (6, -1) is a solution

11. 2(0) - 3(0) = 0, 0 < 4, (0, 0) is not a solution

   2(1) - 3(-1) = 5, 5 > 4, (1, -1) is a solution

   2(3) - 3(0) = 6, 6 > 4, (3, 0) is a solution

13. 3 > -2, (-2, 3) is not a solution

   0 > -2, (-2, 0) is not a solution

   -2 ≤ -2, (0, -2) is a solution

15. For x-intercept;

$0 = -3x + 2, \ 3x = 2, \ x = \frac{2}{3}$

For y-intercept; $y = 2$

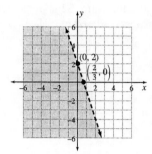

17. For x-intercept;

$0 = 4x - 1, 4x = 1, \ x = \frac{1}{4}$

For y-intercept; $y = -1$

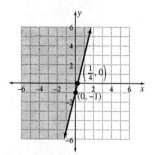

19. For x-intercept;

$0 = 2 - 2x, 2x = 2, x = 1$

For y-intercept; $y = 2$

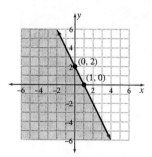

21. For x-intercept;

$2x + 0 = 5, 2x = 5, \ x = \frac{5}{2}$

For y-intercept; $y = 5$

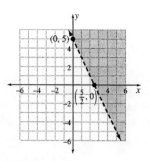

23. For x-intercept;

$2 - 2(0) = 4x, 4x = 2, \ x = \frac{1}{2}$

For y-intercept;

$2 - 2y = 4(0), \ 2 = 2y, \ y = 1$

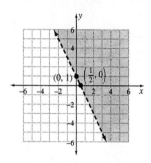

25. x-intercept = 4, no y-intercept

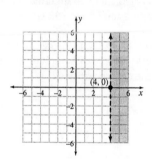

27. No x-intercept, y-intercept = 4

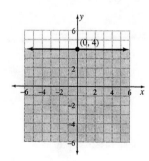

29. Boundary line is $y = 3$

Inequality is $y < 3$

31. Boundary line is $x = 3$

Inequality is $x \leq 3$

33. Boundary line is $y = x + 4$
Inequality is $y \le x + 4$

35. Boundary line is $y = 2x + 1$
Inequality is $y > 2x + 1$

37. Test an ordered pair (that is not on the boundary line) in the inequality, if true shade the side containing the point, if false shade the other side.

39. The first is a line graph with a dot at $x = 3$ and an arrow to the right. The second has a vertical boundary line $x = 3$ and shading to the right.

41. $16x + 12y \ge 2400$

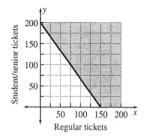

43. a. Could include 0 apricots and 4 tangerines $(0, 4)$; 7 apricots and 0 tangerines $(7, 0)$; 3 apricots and 2 tangerines $(3, 2)$.

b. The region lies in the first quadrant below the line $20a + 35t = 140$, including the line and the coordinate axes.

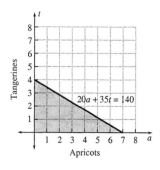

c. $a \ge 0, t \ge 0$
$20a + 35t \le 140$, $a \le 7$, $t \le 4$

45. a. 4 quarters = $1, \$2 - \$1 = \$1$
$\$1 = 10$ dimes, ordered pair is $(10, 4)$

With 0 quarters the entire $2 is in dimes.
$\$2 = 20$ dimes, ordered pair is $(20, 0)$

b. 15 dimes = $1.50,
$\$2 - \$1.50 = \$0.50 = 2$ quarters, ordered pair is $(15, 2)$

With 0 dimes the entire $2 is in quarters.
$\$2 = 8$ quarters, ordered pair is $(0, 8)$

c. $0.10x$

d. $0.25y$

e. $0.10x + 0.25y \le 2$

f. Domain: $0 \le x \le 20$
Range: $0 \le y \le 8$
x and y are both integers.

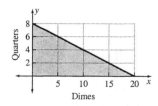

47. 4th quadrant including positive x-axis.

49. 3rd quadrant.

51. $x > 0, y < 0$

Chapter 4 Review Exercises

1. $6 = 4(-1) + b$
$6 = -4 + b$, $b = 10$

3. $37 = \frac{5}{9}(F\text{-}32)$
$\frac{9}{5}(37) = F\text{-}32$, $66.6 + 32 = F$
$F = 98.6$

5. Area in terms of base and height.

7. $W = hp$,
$\dfrac{W}{p} = \dfrac{hp}{p}$,
$h = \dfrac{W}{p}$

9. $A = \dfrac{bh}{2}$,

$2A = 2 \cdot \dfrac{bh}{2}$,

$\dfrac{2A}{h} = \dfrac{bh}{h}$,

$b = \dfrac{2A}{h}$

11. $I = \dfrac{AH}{T}$,

$I \cdot T = T \cdot \dfrac{AH}{T}$,

$\dfrac{IT}{I} = \dfrac{AH}{I}$,

$T = \dfrac{AH}{I}$

13. $PV = nRT$,

$\dfrac{PV}{nR} = \dfrac{nRT}{nR}$,

$T = \dfrac{PV}{nR}$

15. $ax + by = c$,
$ax = c - by$

$x = \dfrac{c - by}{a}$

17. $C = 35 + 5(k - 100)$,
$C - 35 = 5(k - 100)$

$\dfrac{C}{5} - 7 = k - 100$,

$k = \dfrac{C}{5} + 93$

19. $P_1 V_1 = P_2 V_2$,

$\dfrac{P_1 V_1}{V_2} = \dfrac{P_2 V_2}{V_2}$,

$P_2 = \dfrac{P_1 V_1}{V_2}$

21. $A = \frac{1}{2} h (b_1 + b_2)$,

$2A = h(b_1 + b_2)$,

$\dfrac{2A}{h} = b_1 + b_2$,

$b_1 = \dfrac{2A}{h} - b_2$

23. $27 = \frac{1}{2}(3)(9 + b)$

Using solution to exercise #21:

$b = \dfrac{2(27)}{3} - 9 = 18 - 9 = 9$

25. $f(x) = 7 - 2x$

$f(0) = 7 - 2(0) = 7$

$f(3) = 7 - 2(3) = 7 - 6 = 1$

$f(-5) = 7 - 2(-5) = 7 + 10 = 17$

$f(a) = 7 - 2a$

27. $f(x) = x^2 + x$

$f(0) = 0^2 + 0 = 0$

$f(3) = 3^2 + 3 = 9 + 3 = 12$

$f(-5) = (-5)^2 + (-5) = 25 - 5 = 20$

$f(a) = a^2 + a$

29. a. Inputs are x values, $\{2, 4, 6, 8\}$

   b. Outputs are y values
$\{25, 50, 75, 100\}$

   c. The value of "bits" in cents.

31. a. Domain is x values,
$\{1, 2, 3, 4, 5, 6\}$

   b. Range is y values,
$\{2, 6, 10, 20, 40, 60\}$

   c. The name of the nail in "pennies", usually abbreviated d.

33. One output for each input, describes a function.

35. One output for each input, describes a function.

37. Many outputs for same input, not a function.

39.

	Function	Sketch	Domain	Range
a.	$f(x) = x^2$	$y = x^2$	All real numbers, $\mathbb{R}$	$y \geq 0$
b.	$f(x) = x$	$y = x$	All real numbers, $\mathbb{R}$	All real numbers, $\mathbb{R}$
c.	$f(x) = 2$	$y = 2$	All real numbers, $\mathbb{R}$	$y = 2$

41. a. $\dfrac{4-(-2)}{6-4} = \dfrac{6}{2} = 3$

b. $\dfrac{3-(-4)}{-6-4} = \dfrac{7}{-10} = -\dfrac{7}{10}$

c. $\dfrac{0-(-2)}{6-(-4)} = \dfrac{2}{10} = \dfrac{1}{5}$

d. $\dfrac{1-3}{-6-9} = \dfrac{-2}{-15} = \dfrac{2}{15}$

NOTE: The following graph is for #43 and #45

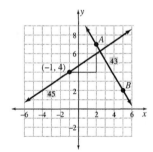

43. Answers may vary. One possibility is to plot the line containing the points $A(2,7)$ and $B(5,2)$. (see graph above)

45. (See graph above)

47. Line goes down from left to right.

49. Write the slope in the form $\dfrac{a}{b}$ where $b > 0$.

Locate (x, y), count a units up (or down) and b units to the right and make a second point. Draw the line between the points.

51. f(0) is the y-intercept.

53. b. Substitute y = 0 into the equation and solve for x.

55. a. Let x = a.

57. f(x) = 0,
2x - 5 = 0,
2x = 5,
$x = \dfrac{5}{2} = 2\dfrac{1}{2}$

59. f(x) = 0
-3x - 4 = 0,
-3x = 4,
$x = -\dfrac{4}{3} = -1\dfrac{1}{3}$

61. f(x) = 0,
$\dfrac{1}{2}x - 6 = 0$,
$\dfrac{1}{2}x = 6$,
x = 12

63. 3x + 5y = 15,
5y = -3x + 15
$y = -\dfrac{3}{5}x + 3$,
$m = -\dfrac{3}{5}, \quad b = 3$

65. $5x - 2y = 10$,
$-2y = -5x + 10$
$y = \frac{5}{2}x - 5$,
$m = \frac{5}{2}$, $b = -5$

67. $4y - 3x = 8$,
$4y = 3x + 8$
$y = \frac{3}{4}x + 2$,
$m = \frac{3}{4}$, $b = 2$

69. $y + 3 = 0$
$y = -3$,
$m = 0$, $b = -3$

71. $x = 3$,
Slope is undefined, there is no y-intercept.

73. $c = 2 + 1.5(n - 1)$
$c = 2 + 1.5n - 1.5$
$c = 1.5n + 0.5$
$m = 1.5$, $b = 0.5$

NOTE: The following graph is for #75-81 (odd)

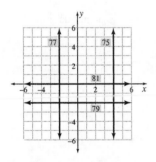

75. $x = 4$
(See graph above)

77. $x = -2$
(See graph above)

79. $y = -2$
(See graph above)

81. $y = 0$ (x-axis)
(See graph above)

NOTE: The following graph is for #83-89 (odd)

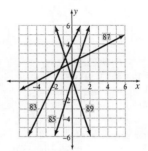

83. $y = 2x + 4$
(See graph above)

85. $y = 3x$
(See graph above)

87. $y = \frac{1}{2}x + 2$, $2y = x + 4$
(See graph above)

89. $y = -3x + \frac{1}{4}$, $4y = -12x + 1$
(See graph above)

91. Rise = -1, run = 2; $m = -\frac{1}{2}$, $b = 1$
$y = -\frac{1}{2}x + 1$

93. Δx is 2 between each input, Δy is -3 between each output, linear function. Slope $m = \frac{-3}{2}$.

95. Δx is 2 between each input, Δy varies, non-linear function.

97. a. Let x = minutes and y = cost
$(14, 4.44)$, $(5, 2.10)$

b. $m = \dfrac{4.44 - 2.10}{14 - 5} = \dfrac{2.34}{9} = 0.26$

c. Slope is cost per minute.

d. $2.10 = (0.26)(5) + b$
$2.10 = 1.3 + b$
$0.80 = b$
$y = 0.26x + 0.8$

99. a. Let x = hours and y = feet
$(12, 12)$, $(17, 32)$

b. $m = \dfrac{32 - 12}{17 - 12} = \dfrac{20}{5} = 4$

c. Slope is feet per hour.

d. $32 = 4(17) + b$

$32 = 68 + b$

$-36 = b$

$y = 4x - 36$

101. (d); Solving y + b = mx for y gives
y = mx - b, not equivalent to the other
equations.

103. Parallel line has same slope, $b = 0$
$y = 3x$

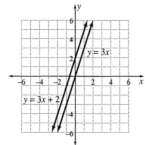

105. Perpendicular line has slope $m = \frac{3}{2}$,

$b = -2$. $y = \frac{3}{2}x - 2$

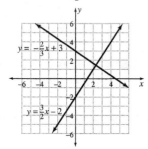

107. $C = 15w - 50$

109. $C = 20w - 150$

111. a. x, miles

b. y, cost in dollars

c. yes.

d. From the table, 125 is between 100 & 150,
cost will be between 43 & 47. Extending
the table, the cost of driving 125 miles is
$45.

e. From the table it appears that every 25
miles costs an additional $2. Extending
the table, for $57 you can drive 275 miles.

f. The y-intercept is $35. This is the base
cost for the rental.

g. $\Delta x = 50$, $\Delta y = 4$

slope $= \dfrac{4}{50} = 0.08$ or 0.08

This represents the cost per mile.

h. $m = 0.08$, $b = 35$

$y = \$0.08x + \$35$

i. $0 = 0.08x + 35$

$-0.08x = 35$

$x = -437.5$

This has no meaning in the context of the
problem.

113. $-6 > 3(0) - 6$ $-12 > 3(-2) - 6$

$-6 > -6$ false $-12 > -12$ false

$1 > 3(2) - 6$ $4 > 3(3) - 6$

$1 > 0$ true $4 > 3$ true

Ordered pairs (c) and (d) make the inequality
true.

115. a. All points on the y-axis or in quadrant 2 or
3.

b. All points above the x-axis.

117. Boundary equation: $y = 3x - 2$

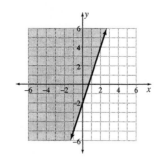

119. Boundary equation: $y = 4x + 10$

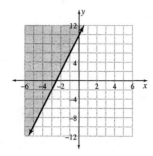

121. Boundary equation: $y = -\dfrac{2}{3}x + 3$

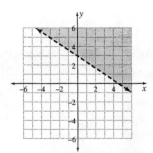

Chapter 4 Test

1. $b = 5 - \frac{1}{2}(-8),$

$b = 9$

2. Answers will vary but could include:

$G = \dfrac{T_1 + T_2 + H + F}{\text{Total Points}}$

3. a. $C = \pi d,$

$\dfrac{C}{\pi} = \dfrac{\pi d}{\pi},$

$d = \dfrac{C}{\pi}$

b. $A = \frac{1}{2}bh,$

$2A = bh,$

$h = \dfrac{2A}{b}$

c. $y = mx + b,$

$b = y - mx$

d. $P_1 V_1 = P_2 V_2,$

$\dfrac{P_1 V_1}{P_2} = \dfrac{P_2 V_2}{P_2},$

$V_2 = \dfrac{P_1 V_1}{P_2}$

4. a. One output for each input, function.

b. Many outputs for one input, not a function.

c. One output for each input, function.

d. One output for each input, function.

e. Many outputs for one input, not a function.

f. One output for each input, function.

5. a. (1) Horizontal line, $m = 0$

(2) Rise = 50, run = 5, $m = \frac{50}{5} = 10$

(3) Rise = 300, run = 6,

$m = \frac{300}{6} = 50$

(4) Vertical line, slope is undefined

b. (2), slope = 10

6. a. $m = \dfrac{-1 - (-3)}{2 - 5} = \dfrac{2}{-3} = -\dfrac{2}{3}$

b. $m = \dfrac{-4 - (-4)}{0 - 1} = \dfrac{0}{-1} = 0$

c. $m = \dfrac{-4 - 3}{2 - 1} = \dfrac{-7}{1} = -7$

d. $m = \dfrac{1 - (-3)}{4 - 4} = \dfrac{4}{0},$ *undefined*

7. a. $y = 5 - 2x,\ m = -2,\ b = 5$

b. $6y + 3 = 2x,$

$6y = 2x - 3$

$y = \frac{2}{6}x - \frac{3}{6},$

$y = \frac{1}{3}x - \frac{1}{2},\ m = \frac{1}{3},\ b = -\frac{1}{2}$

8.

9.

10. a. $f(x) = 0,\ (y = 0$ is the x-axis)

b. $f(0),\ (x = 0$ is the y-axis)

    c.   f(a)

11. y = 5x - 1

12. $y = \frac{1}{3}x + 2, \quad 3y = x + 6$

13. **a.** Different slopes, same y-intercept, neither.

    b. Same slope, different y-intercept, parallel.

    c. Slopes are opposite reciprocals, perpendicular.

    d. Reciprocal slopes, different y-intercepts, neither

14. Substitute slope and ordered pair into b = y - mx and solve for b. Put b and m into y = mx + b.

15. Parallel line has same slope, *a* and given y-intercept, *k*. y = ax + k.

16. For a horizontal line y = k.

17.

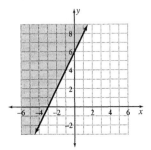

18.

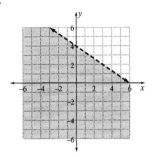

19. **a.** x, Miles

    b. y, Cost

    c. Yes.

    d. From the table,125 is between 100 & 150, cost will be between 40 & 45. Extending the table, the cost is $42.50.

    e. From the table it appears that every 10 miles costs an additional $1. Extending the table, $57 will buy 270 miles.

    f. $30; Base price to rent the vehicle.

    g. $\Delta x = 50, \Delta y = 5$, slope $= \frac{5}{50} = 0.10$, the cost per mile.

    h. $y = 0.10x + 30$

    i. 0 = 0.10x + 30, -30 = 0.10x
       x = -300, has no meaning.

Cumulative Review of Chapters 1 to 4

1. $\frac{1}{3} + \frac{3}{7} = \frac{7}{21} + \frac{9}{21} = \frac{16}{21}$

    $\frac{1}{3} - \frac{3}{7} = \frac{7}{21} - \frac{9}{21} = -\frac{2}{21}$

    $\frac{1}{3} \cdot \frac{3}{7} = \frac{3}{21} = \frac{1}{7}$

    $\frac{1}{3} \div \frac{3}{7} = \frac{1}{3} \cdot \frac{7}{3} = \frac{7}{9}$

3. 4.8 + (-6.4) = 4.8 - 6.4 = -1.6
    4.8 - (-6.4) = 4.8 + 6.4 = 11.2
    4.8(-6.4) = -30.72
    4.8 ÷ (-6.4) = -0.75

5. $4x^2 + 6x$
    $4x^2 - 6x$
    $4x^2(6x) = (4)(6)(x^2)(x) = 24x^3$
    $4x^2 \div 6x = \dfrac{4x^2}{6x} = \frac{2}{3}x$

7. $1,000,000 \text{ min} \cdot \dfrac{1 \text{ hr}}{60 \text{ min}} \cdot \dfrac{1 \text{ day}}{24 \text{ hr}}$
    ≈ 694.4 days

9. $A = 91 \text{ in}^2$, (a + b) = 13 in
    $91 = \frac{1}{2}h(13)$,
    182 = 13h,
    h = 14 in

11. 2(x + 5) = x + 3
    2x + 10 = x + 3, x + 10 = 3
    x = -7

13. 9 - 2(x - 3) = 21, -2(x - 3) = 12
    x - 3 = -6, x = -3

15. **a.** $m = -\frac{3}{2}$

    b. $b = 9$

17.

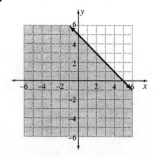

19. $f(0)$ is the vertical axis intercept.

21. $f(x) = x + 6$
$f(-2) = -2 + 6$
$ = 4$

23. $s = \dfrac{a}{1-r}$

$s(1-r) = \dfrac{a}{1-r} \cdot (1-r)$

$s(1-r) = a$

       or

$a = s(1-r)$

25. Choosing the points $(0,2)$ and $(1,0)$, we have
$\Delta x = 1$ and $\Delta y = -2$. Thus,

$m = \dfrac{\Delta y}{\Delta x} = \dfrac{-2}{1} = -2$.

The y-intercept is $b = 2$.
$y = mx + b$
$y = -2x + 2$

Chapter 5

Exercises 5.1

1. a. $\frac{5}{3}$, 5:3

   b. 3 : 2, 3 to 2

   c. $\frac{4}{9}$, 4 to 9

3. a. Common factor is 5
   $$\frac{15}{35} = \frac{5 \cdot 3}{5 \cdot 7} = \frac{3}{7}$$

   b. Common factor is 16
   $$\frac{48}{16} = \frac{16 \cdot 3}{16 \cdot 1} = \frac{3}{1}$$

   c. Common factor is 240
   $$\frac{5280}{3600} = \frac{240 \cdot 22}{240 \cdot 15} = \frac{22}{15}$$

5. a. $\dfrac{3}{15} = \dfrac{3 \cdot 1}{3 \cdot 5} = \dfrac{1}{5}$

   b. $\dfrac{7.5}{10} = \dfrac{2.5 \cdot 3}{2.5 \cdot 4} = \dfrac{3}{4}$

   c. $\dfrac{6.75}{10.5} = \dfrac{0.75 \cdot 9}{0.75 \cdot 14} = \dfrac{9}{14}$

   d. $\dfrac{7.25}{9.75} = \dfrac{0.25 \cdot 29}{0.25 \cdot 39} = \dfrac{29}{39}$

7. a. Slope = 3, slope ratio is $\frac{3}{1}$.

   b. Slope = 2 , slope ratio is $\frac{2}{1}$.

   c. Slope = -4, slope ratio is $\frac{-4}{1}$.

9. a. Slope = 2.98, slope ratio = $\frac{2.98}{1}$ or $\frac{149}{50}$

   b. Slope = -1.50 ,
   slope ratio is $\dfrac{-1.50}{2}$ or $-\dfrac{3}{2}$.

11. a. $\frac{1}{2}$ foot $\cdot \dfrac{12 \text{ inches}}{1 \text{ foot}}$ = 6 inches,
   ratio is 6 to 1

   b. 3000 grams $\cdot \dfrac{1 \text{ kilogram}}{1000 \text{ grams}} = 3 \text{ kg}$
   ratio is 3 to 1

   c. 32 ounces $\cdot \dfrac{1 \text{ pound}}{16 \text{ ounces}} = 2 \text{ pounds}$
   ratio is 2 to 5

   d. $2\frac{2}{3}$ yards $\cdot \dfrac{3 \text{ feet}}{1 \text{ yard}} = 8 \text{ feet}$
   ratio 8 to 2 = 4 to 1

   e. 20 liters $\cdot \dfrac{1000 \text{ ml}}{1 \text{ liter}} = 20,000 \text{ ml}$
   ratio 200 to 20,000 = 1 to 100

   f. 2 years = 24 months
   ratio 24 to 150 = 4 to 25

   g. $\frac{1}{4}$ hour = 15 minutes
   ratio 20 to 15 = 4 to 3

13. 3 to 1

15. $\frac{1}{2}$ pint $\cdot \dfrac{1 \text{ qt}}{2 \text{ pt}} \cdot \dfrac{1 \text{ gal}}{4 \text{ qt}} = \dfrac{1}{16}$ gal
   The ratio is 1 to 16.

17. $\dfrac{240 \text{ cc}}{16 \text{ hours}} = 15$ cc per hour

19. a. Common factors are 4x,
   $12x : 4x^2 = 3 : x$

   b. Common factors are 2xy,
   $2xy$ to $6x^2y$ = 1 to 3x

   c. Common factor is x,
   $x : 3x^4 = 1 : 3x^3$

21. a. Common factor is 6.
   $24(x+1) : 6 = 4(x+2) : 1$
   $= 4x + 8 : 1$

   b. Common factor is 3.
   $15(x-1) : 3 = 5(x-1) : 1$
   $= 5x - 5 : 1$

   c. Common factor is 4.
   $12(x+3) : 4 = 3(x+3) : 1$
   $= 3x + 9 : 1$

23. Answers will vary

25. (b), (c), and (d) are correct entries.
The entry in (a) would give
$$\frac{60 \cdot 60 \cdot 3 \cdot 16}{16}$$

27. a. $\dfrac{30 \text{ miles}}{\text{hour}} \cdot \dfrac{5280 \text{ feet}}{1 \text{ mile}} \cdot \dfrac{1 \text{ hour}}{60 \text{ min}} \cdot \dfrac{1 \text{ min}}{60 \text{ sec}}$
= 44 feet per second

b. $\dfrac{55 \text{ miles}}{1 \text{ hr}} \cdot \dfrac{1 \text{ hr}}{3600 \text{ sec}} \cdot \dfrac{5280 \text{ ft}}{1 \text{ mi}}$
$= \dfrac{290,400}{3600}$ ft/sec
$= \dfrac{242}{3}$ ft/sec
$= 80\dfrac{2}{3}$ ft/sec

29. a. $\dfrac{88 \text{ ft}}{\text{sec}} \cdot \dfrac{60 \text{ sec}}{1 \text{ min}} \cdot \dfrac{60 \text{ min}}{1 \text{ hour}} \cdot \dfrac{1 \text{ mile}}{5280 \text{ ft}}$
= 60 miles per hour

b. $\dfrac{66 \text{ ft}}{1 \text{ sec}} \cdot \dfrac{1 \text{ mi}}{5280 \text{ ft}} \cdot \dfrac{3600 \text{ sec}}{1 \text{ hr}}$
$= \dfrac{237,600}{5280}$ mi/hr
= 45 miles per hour

31. $\dfrac{4 \text{ sec}}{1} \cdot \dfrac{60 \text{ miles}}{\text{hour}} \cdot \dfrac{1 \text{ hour}}{60 \text{ min}}$
$\cdot \dfrac{1 \text{ min}}{60 \text{ sec}} \cdot \dfrac{5280 \text{ ft}}{1 \text{ mile}}$
= 352 feet, more than a city block

33. $\dfrac{25 \text{ hours}}{1 \text{ oil change}} \cdot \dfrac{100 \text{ feet}}{30 \text{ sec.}} \cdot \dfrac{1 \text{ mile}}{5280 \text{ ft}} \cdot \dfrac{3600 \text{ sec}}{1 \text{ hr}}$
≈ 56.8 miles per oil change

35. $\dfrac{30 \text{ min}}{1} \cdot \dfrac{6 \text{ miles}}{1 \text{ hour}} \cdot \dfrac{1 \text{ hour}}{60 \text{ min}} = 3$ miles

37. $\dfrac{250 \text{ mL}}{12 \text{ hours}} \cdot \dfrac{60 \text{ microdrops}}{1 \text{ mL}} \cdot \dfrac{1 \text{ hr}}{60 \text{ min}}$
= 20.8 microdrops per minute

39. $62.5\% = \dfrac{62.5}{100} = \dfrac{5 \cdot 12.5}{8 \cdot 12.5} = \dfrac{5}{8}$

41. $\dfrac{1}{12} = 0.08\overline{33} = 8\tfrac{1}{3}\%$

43. $\dfrac{1}{8} = 12.5\%$

45. 1 mile = 5280 feet
30 feet per mile $= \dfrac{30}{5280}$
≈ 0.006
$= 0.6\%$

47. $\dfrac{19.99 - 12}{12} = \dfrac{7.99}{12} = 0.6658$
The percent change is roughly 66.6%.
The 2003 price is roughly 166.6% of the 1986 price.

49. $\dfrac{10.99 - 5}{5} = \dfrac{5.99}{5} = 1.198$
The percent change is 119.8%. The 2003 price is 219.8% of the 1986 price.

51. a. $15 + 15(0.25) = 15 + 3.75 = 18.75$
The ending value is $18.75.

b. $18.75 - 18.75(0.25) \approx 18.75 - 4.69$
$\qquad = 14.06$
The sale price of the belt is $14.06.

53. a. $12 + 12(0.50) = 12 + 6 = 18$
The puppy grows to be 18 inches tall.

b. $18 - 18(0.50) = 18 - 9 = 9$
The DVD is on sale for $9.

55. The original values are different. The percent decrease is based on a larger number.

57. a. Let x be the new rent;
$$\frac{x - 450}{450} = 0.20, \quad x - 450 = 90$$
x = $540

b. Let x be the old rent;
$$\frac{450 - x}{x} = 0.20, \quad 450 - x = 0.20x$$
$450 = 1.20x, \; x = \$375$

59. Total income is $1200 (sum of all expenditures)

$$\frac{360}{1200} = 30\%, \quad \frac{132}{1200} = 11\%, \quad \frac{240}{1200} = 20\%,$$

$$\frac{84}{1200} = 7\%, \quad \frac{96}{1200} = 8\%, \quad \frac{48}{1200} = 4\%,$$

$$\frac{60}{1200} = 5\%, \quad \frac{24}{1200} = 2\%, \quad \frac{156}{1200} = 13\%$$

61. Time on A : Time on B : Time on C

4 : 3 : 1

$$400\left(\frac{4}{4+3+1}\right) = 400\left(\frac{1}{2}\right) = 200$$

$$400\left(\frac{3}{4+3+1}\right) = 400\left(\frac{3}{8}\right) = 150$$

$$400\left(\frac{1}{4+3+1}\right) = 400\left(\frac{1}{8}\right) = 50$$

If a total of $400 is billed for the day, then $200 is charged to project A, $150 is charged to project B, and $50 is charged to project C.

63. Continued ratio of 4 : 3 : 3 is a total of 10 shares, $2500 \div 10 = 250$ per share,

Cosmo gets $4(250) = 1000$ crowns

Timo and Sven get $3(250) = 750$ crowns each.

65. a. 9 carbon : 8 hydrogen : 4 oxygen

   b. 1 sodium :1 hydrogen : 1 carbon : 3 oxygen.

   c. 1 calcium : 1 carbon : 3 oxygen

   d. 12 carbon : 22 hydrogen : 11 oxygen

   e. 8 carbon : 10 hydrogen : 4 nitrogen : 2 oxygen

67. a. $209 \div 585 = 0.357$

   b. $152 \div 584 = 0.260$

   c. $174 \div 565 \approx 0.308$

   d. $141 \div 470 = 0.300$

   e. $236 \div 539 \approx 0.438$

   f. $227 \div 536 \approx 0.424$

69. Divide and compare the decimals.

$$\frac{5}{12} \approx 0.417$$

$$\frac{7.5}{10.5} \approx 0.714$$

$$\frac{6.75}{10.75} \approx 0.628$$

The slopes in order from flattest to steepest are:

$$\frac{5}{12}, \quad \frac{6.75}{10.75}, \quad \frac{7.5}{10.5}$$

Exercises 5.2

1. 3 : 1, unsafe, slips

3. 4.5 : 1, unsafe, tips

5. 4 : 1, safe

7. $\frac{6}{8} \overset{?}{=} \frac{15}{20}, \quad 6(20) \overset{?}{=} 8(15), \quad 120 = 120$

proportion

9. $\frac{4}{6} \overset{?}{=} \frac{6}{9}, \quad 4(9) \overset{?}{=} 6(6), \quad 36 = 36$

proportion

11. $\frac{9}{21} \overset{?}{=} \frac{21}{35}, \quad 9(35) \overset{?}{=} 21(21), \quad 315 \neq 441$

false statement

13. $\frac{2}{4} = \frac{5}{10}, \quad 2(10) = 4(5), \quad 20 = 20$

$\frac{4}{2} = \frac{10}{5}, \quad 4(5) = 2(10), \quad 20 = 20$

$\frac{10}{4} = \frac{5}{2}, \quad 10(2) = 4(5), \quad 20 = 20$

$\frac{4}{10} = \frac{2}{5}, \quad 4(5) = 10(2), \quad 20 = 20$

15. $\frac{3}{4} = \frac{x}{15}, \quad 4x = 15 \cdot 3, \quad x = \frac{15 \cdot 3}{4},$

$x = 11\frac{1}{4}$

17. $\frac{x}{5} = \frac{2}{3}, \quad 3x = 2 \cdot 5, \quad x = \frac{5 \cdot 2}{3},$

$x = 3\frac{1}{3}$

19. $\dfrac{7}{3} = \dfrac{5}{x}$, $7x = 3 \cdot 5$, $x = \dfrac{3 \cdot 5}{7}$,

    $x = 2\dfrac{1}{7}$

21. $\dfrac{4}{x} = \dfrac{3}{7}$, $3x = 7 \cdot 4$, $x = \dfrac{7 \cdot 4}{3}$,

    $x = 9\dfrac{1}{3}$

23. $0.45x = 36$, $x = \dfrac{36}{0.45}$, $x = 80$

    or $\dfrac{45}{100} = \dfrac{36}{x}$, $45x = 36(100)$,

    $x = \dfrac{3600}{45}$, $x = 80$

25. $56 = \dfrac{x}{100} \cdot 84$, $x = \dfrac{5600}{84}$, $x \approx 66.7\%$

    or $\dfrac{56}{84} = \dfrac{x}{100}$, $84x = 56(100)$,

    $x = \dfrac{5600}{84}$, $x \approx 66.7\%$

27. $\dfrac{60}{100} = \dfrac{x}{42}$, $100x = 60(42)$,

    $x = \dfrac{2520}{100}$, $x = 25.2$

29. $\dfrac{x}{100} = \dfrac{56}{40}$, $40x = 5600$,

    $x = \dfrac{5600}{40}$, $x = 140\%$

31. $\dfrac{18}{100} = \dfrac{x}{25}$, $100x = 450$,

    $x = \dfrac{450}{100}$, $x = 4.5$

33. $0.64x = 16$, $x = \dfrac{16}{0.64}$, $x = 25$

    or $\dfrac{64}{100} = \dfrac{16}{x}$, $64x = 16(100)$,

    $x = \dfrac{1600}{64}$, $x = 25$

35. $\dfrac{x}{100} = \dfrac{104}{80}$, $80x = 104(100)$,

    $x = \dfrac{10400}{80}$, $x = 130\%$

37. $\dfrac{x}{100} = \dfrac{28}{40}$, $40x = 28(100)$, $x = \dfrac{2800}{40}$,

    $x = 70\%$

39. $\dfrac{x}{65 \text{ in}} = \dfrac{1 \text{ m}}{39.37 \text{ in}}$, $39.37x \text{ in} = 65 \text{ m in}$,

    $x = \dfrac{65 \text{ m in}}{39.37 \text{ in}}$, $x \approx 1.65 \text{ m}$

41. $\dfrac{x}{15 \text{ mi}} = \dfrac{1 \text{ km}}{0.621 \text{ mi}}$, $0.621x \text{ mi} = 15 \text{ km mi}$,

    $x = \dfrac{15 \text{ km mi}}{0.621 \text{ mi}}$, $x \approx 24.15 \text{ km}$

43. $640 \text{ acres} \cdot \dfrac{43,560 \text{ ft}^2}{1 \text{ acres}}$

    $= 27,878,400 \text{ ft}^2$

45. $\dfrac{x}{6.5 \text{ ft}} = \dfrac{4}{1}$, $x = 4(6.5 \text{ ft})$, $x = 26 \text{ ft}$

47. $54 \text{ in} = 4.5 \text{ ft}$, $\dfrac{4.5 \text{ ft}}{x} = \dfrac{1}{8}$, $x = 36 \text{ ft}$

49. $\dfrac{8 \text{ ft}}{x} = \dfrac{7}{11}$, $7x = 88 \text{ ft}$, $x = \dfrac{88 \text{ ft}}{7}$,

    $x \approx 12.57 \text{ ft}$

51. $\dfrac{13 \text{ miles}}{x} = \dfrac{1 \text{ mile}}{5280 \text{ ft}}$, $x = 68,640 \text{ ft}$

    slope $= \dfrac{\text{rise}}{\text{run}} = \dfrac{10 \text{ ft}}{68640 \text{ ft}} = \dfrac{1}{6864}$

53. $\dfrac{x}{40 \text{ lb}} = \dfrac{500,000 \text{ units}}{150 \text{ lb}}$,

    $150x \text{ lb} = 20,000,000 \text{ units lb}$,

    $x \approx 130,000 \text{ units}$

55. $\dfrac{62.5}{100} = \dfrac{x}{3000}$, $100x = 187,500$

    $x = 1875 \text{ ft}$

57. $\dfrac{9}{100} = \dfrac{x}{2}$, $100x = 18$, $x = \dfrac{18}{100} \text{ mi}$

    $0.18 \text{ mi} \cdot \dfrac{5280 \text{ ft}}{1 \text{ mi}} \approx 950 \text{ ft}$

59. $\dfrac{75 \text{ tagged}}{x} = \dfrac{10 \text{ tagged}}{65 \text{ total birds}}$,

    $10x = 4875$, $x \approx 487$ to 488 total birds

61. a. $30 \cdot 75 = 2250 \text{ ft}^2$

$$\frac{x}{2250 \text{ ft}^2} = \frac{18 \text{ bats}}{1 \text{ ft}^2}$$

$$x = 18(2250) \text{ bats}$$

$$x = 40,500 \text{ bats}$$

b. $$\frac{x}{30 \text{ nights}} = \frac{500 \text{ mosquitoes}}{1 \text{ night}}$$

$$x = 500(30) \text{ mosquitoes}$$

$$x = 15,000 \text{ mosquitoes}$$

Each bat can eat 15,000 mosquitoes in 30 nights. Thus, the total number of mosquitoes eaten in 30 nights by all the bats is

$(40,500)(15,000) = 607,500,000$.

63. For $\dfrac{a}{b} = \dfrac{c}{d}$, cross multiply to get $ad = bc$.

65. Divide x by the number n and set it equal to 20% written as a fraction.

$$\frac{20}{100} = \frac{x}{n}$$

67. Look for a single fraction on each side of the equal sign.

$$\frac{a}{b} = \frac{c}{d}$$

69. $18\left(\dfrac{2x}{9}\right) = 18\left(\dfrac{x+2}{6}\right)$

$$2(2x) = 3(x+2)$$

$$4x = 3x + 6$$

$$x = 6$$

$$\frac{2x}{9} = \frac{x+2}{6}$$

$$6(2x) = 9(x+2)$$

$$12x = 9x + 18$$

$$3x = 18$$

$$x = 6$$

71. $2\left(\dfrac{x}{2}\right) = 2(5x+4)$

$$x = 10x + 8$$

$$-9x = 8$$

$$x = -\frac{8}{9}$$

$$\frac{x}{2} = 5x + 4$$

$$1(x) = 2(5x+4)$$

$$x = 10x + 8$$

$$-9x = 8$$

$$x = -\frac{8}{9}$$

73. $6\left(\dfrac{1}{2} + \dfrac{x}{3}\right) = 6\left(\dfrac{5x}{2} + \dfrac{4}{3}\right)$

$$3 + 2x = 15x + 8$$

$$-13x = 5$$

$$x = -\frac{5}{13}$$

$$\frac{1}{2} + \frac{x}{3} = \frac{5x}{2} + \frac{4}{3}$$

$$\frac{3 + 2x}{6} = \frac{15x + 8}{6}$$

$$6(3 + 2x) = 6(15x + 8)$$

$$18 + 12x = 90x + 48$$

$$-78x = 30$$

$$x = -\frac{30}{78}$$

$$x = -\frac{5}{13}$$

75. $\dfrac{x-1}{2} = \dfrac{x+3}{3}, \quad 3(x-1) = 2(x+3),$

$3x - 3 = 2x + 6, \quad x - 3 = 6, \quad x = 9$

77. $\dfrac{x+4}{15} = \dfrac{x+2}{10}, \quad 10(x+4) = 15(x+2),$

$10x + 40 = 15x + 30, \quad 40 = 5x + 30,$

$10 = 5x, \quad x = 2$

79. $\dfrac{x+7}{9} = \dfrac{x+3}{6}, \quad 6(x+7) = 9(x+3),$

$6x + 42 = 9x + 27, \quad 42 = 3x + 27,$

$15 = 3x, \quad x = 5$

81. $\dfrac{3x}{2} = \dfrac{6x+3}{5}$, $15x = 2(6x+3)$,

     $15x = 12x+6$, $3x = 6$, $x = 2$

83. $\dfrac{2x-3}{7} = \dfrac{x+2}{4}$, $4(2x-3) = 7(x+2)$,

     $8x-12 = 7x+14$, $x-12 = 14$, $x = 26$

85. $\dfrac{4x+2}{5} = \dfrac{x-1}{2}$, $2(4x+2) = 5(x-1)$,

     $8x+4 = 5x-5$, $3x+4 = -5$,

     $3x = -9$, $x = -3$

87. $\dfrac{x+2}{2} = \dfrac{3x+1}{5}$, $5(x+2) = 2(3x+1)$,

     $5x+10 = 6x+2$, $10 = x+2$, $x = 8$

Exercises 5.3

1. $\dfrac{7.5}{12} = \dfrac{h}{8}$, $12h = 7.5(8)$, $h = \dfrac{7.5 \cdot 8}{12}$,

     $h = 5$ cm

3. Trace the rectangle. Draw a diagonal.
Increase the base to 2.5 inches. Draw a
vertical line at 2.5 inches up to diagonal.
Vertical line is 1.25 inches.

5. Trace the rectangle. Draw a diagonal.
Decrease the base to 1.0 inches. Draw a
vertical line at 1.0 inches up to diagonal.
Vertical line is 0.2 inches.

7. Ratios are $\dfrac{16}{25} = \dfrac{16}{25}$, similar triangles
Corresponding sides:
Sides: AB and ND, AT and NE, BT and DE

9. Ratios are $\dfrac{10}{29}$ and $\dfrac{29}{34}$, $\dfrac{10}{29} \neq \dfrac{29}{34}$,
triangles are not similar

11. Ratios are $\dfrac{6}{12}$ and $\dfrac{16}{32}$, both equal $\dfrac{1}{2}$,
figures are similar.

Corresponding sides are: FG and HK, FR and
HA, RO and AW, GO and KW

13. Ratios are $\dfrac{11}{8}$ and $\dfrac{22}{16}$, $\dfrac{11}{8} = \dfrac{22}{16}$,
figures are similar.
Corresponding segments are: radii RT and
OE, diameters EN and DV

15. $\dfrac{28}{56} = \dfrac{n}{63}$, $56n = 28(63)$, $n = \dfrac{28(63)}{56}$,

     $n = 31.5$

17. $\dfrac{12}{6} = \dfrac{n}{24}$, $6n = 12(24)$, $n = \dfrac{12(24)}{6}$,

     $n = 48$

19. $\dfrac{35}{30} = \dfrac{s}{4}$, $30s = 35(4)$ $s = \dfrac{35(4)}{30}$,

     $s \approx 4.7$ ft

21. $\dfrac{7}{1.4} = \dfrac{t}{4}$, $1.4t = 7(4)$, $t = \dfrac{7(4)}{1.4}$,

     $t = 20$ ft

23. $\dfrac{14}{27} = \dfrac{t}{78}$, $27t = 14(78)$, $t = \dfrac{14(78)}{27}$

     $t \approx 40.4$ ft

25. $\dfrac{8}{5} = \dfrac{4+x}{x}$, $8x = 5(4+x) = 20+5x$

     $3x = 20$, $x = \dfrac{20}{3}$, $x = 6\dfrac{2}{3}$

27. $\dfrac{15-5}{15} = \dfrac{6}{x}$, $10x = 15(6)$,

     $10x = 90$, $x = 9$

29. $\dfrac{4}{3} = \dfrac{12}{12-x}$, $4(12-x) = 3(12)$,

     $48-4x = 36$, $4x = 12$, $x = 3$

31. Let x = distance between origin and A. Using
similar triangles:
$\dfrac{3}{x} = \dfrac{2}{4}$, $2x = 12$, $x = 6$,   A = (6, 0)
B = (3, 2), from points given on graph

33. A = (0, 2), from points given on graph
Let y = distance from origin to B. The
distance from A to B is then y - 2.

Using similar triangles:

$\dfrac{6}{3.5} = \dfrac{y}{y-2}, \quad 3.5y = 6(y-2)$

$3.5y = 6y - 12, \quad 2.5y = 12, \quad y = 4.8$

$B = (0, 4.8)$

35. Think of the parking meter and its shadow as the height and base of a small triangle inside of a larger triangle. The larger triangle has a base of the distance from the street light (x) plus the shadow and a height of the street light.

$\dfrac{5}{4} = \dfrac{15}{x+4}, \quad 5(x+4) = 60,$

$5x + 20 = 60$

$5x = 40, \quad x = 8 \text{ ft}$

37. $AB = 10 - x$

39. $BC = x - 10$

41. To identify similar triangles, check to see if corresponding sides are proportional or corresponding angles are equal.

43. Squares, circles and spheres are always similar.

Mid-Chapter 5 Test

1. $12xy$ to $15y^2z = 3(4)xy$ to $3(5)yyz$

$= 4x$ to $5yz$

2. $\dfrac{24(x-2)}{8} = \dfrac{8 \cdot 3(x-2)}{8}$

$\qquad = \dfrac{3(x-2)}{1}$

$\qquad = \dfrac{3x-6}{1}$

3. 6 feet = 72 inches

ratio is $18 : 72 = 1 : 4$

4. 3 meters = 300 centimeters

ratio is 300 to 75 = 4 to 1

5. $\dfrac{16}{100} = \dfrac{11.2}{x}, \quad 16x = 11.2(100), \quad x = 70$

or $0.16x = 11.2, \quad x = \dfrac{11.2}{0.16}, \quad x = 70$

6. $\dfrac{360}{x} = \dfrac{250}{100}, \quad 250x = 360(100), \quad x = 144$

or $2.50x = 360, \quad x = \dfrac{360}{2.50}, \quad x = 144$

7. $\dfrac{\text{new - original}}{\text{original}} = \dfrac{30,000 - 24,000}{24,000}$

$= \dfrac{6,000}{24,000} = \dfrac{1}{4} = 0.25, \quad 25\% \text{ increase}$

8. $\dfrac{\text{new - original}}{\text{original}} = \dfrac{13.99 - 15.99}{15.99}$

$= \dfrac{-2}{15.99} \approx -0.125, \quad 12.5\% \text{ decrease}$

9. Answers may vary. One possible answer is the fuel usage in an aircraft.

10. Answers may vary. One possible answer is production numbers in a chocolate factory.

11. $\dfrac{5}{x} = \dfrac{500,000}{1}, \quad x = \dfrac{5}{500,000},$

$x = \dfrac{1}{100,000}$

12. $\dfrac{3}{5} = \dfrac{16}{x}, \quad 3x = 5(16), \quad x = \dfrac{80}{3}, \quad x = 26\tfrac{2}{3}$

13. $\dfrac{x}{12} = \dfrac{10}{8}, \quad 8x = 120, \quad x = \dfrac{120}{8}, \quad x = 15$

14. $\dfrac{x+5}{8} = \dfrac{2x-1}{12}, \quad 12(x+5) = 8(2x-1)$

$12x + 60 = 16x - 8, \quad 68 = 4x, \quad x = \dfrac{68}{4},$

$x = 17$

15. $\dfrac{5x-2}{3x} = \dfrac{4}{3}, \quad 3(5x-2) = 4(3x)$

$15x - 6 = 12x, \quad 3x = 6, \quad x = 2$

16. The equation is not a proportion because it is not in the form $\dfrac{a}{b} = \dfrac{c}{d}$.

$x - 3 = \dfrac{13}{3} - \dfrac{1}{3}x$

$x + \dfrac{1}{3}x = \dfrac{13}{3} + 3$

$\dfrac{4}{3}x = \dfrac{22}{3}$

$x = \left(\dfrac{3}{4}\right)\left(\dfrac{22}{3}\right)$

$x = \dfrac{11}{2} \text{ or } x = 5.5$

17. $\dfrac{a}{b} = \dfrac{c}{d}$, $bc = ad$, $b = \dfrac{ad}{c}$

18. $\dfrac{6}{8} = \dfrac{x}{15}$, $8x = 6(15)$, $x = \dfrac{6(15)}{8}$, $x = \dfrac{90}{8}$,
    $x = 11\frac{1}{4}$

19. A = (5, 0), from the points in the figure
    Let y = the height between B and
    (7, 0).
    Using similar triangles:
    $\dfrac{5}{7} = \dfrac{4}{y}$, $5y = 4\cdot7$, $y = \dfrac{28}{5}$, $y = 5.6$
    B = (7, 5.6)

20. 2 + 3 + 5 = 10 total parts,
    1500 ÷ 10 = 150 per part

    fat = 2(150) = 300
    $\dfrac{300}{1500} = \dfrac{1}{5} = 20\%$

    carbohydrate = 3(150) = 450
    $\dfrac{450}{1500} = \dfrac{3}{10} = 30\%$

    protein = 5(150) = 750
    $\dfrac{750}{1500} = \dfrac{1}{2} = 50\%$

    or let x = one part,
    2x + 3x + 5x = 1500,
    10x = 1500, x = 150,
    2x = 2(150) = 300,
    3x = 3(150) = 450,
    5x = 5(150) = 750

21. 1 + 1 + 2 = 4 parts
    180 ÷ 4 = 45 per part
    Angles are $45°, 45°$, and $2(45°) = 90°$
    or let x = 1 part, 1x + 1x + 2x = 180,
    4x = 180, x = 45,
    2x = 2(45) = 90
    Angles are $45°, 45°$ and $90°$

22. $\dfrac{1}{12} = \dfrac{48}{x}$, $x = 48(12)$, $x = 576$ inches

    576 inches $\cdot \dfrac{1\ ft}{12\ inches} = 48$ ft

23. $\dfrac{\text{clipped fin}}{\text{not clipped}} = \dfrac{\text{clipped fin}}{\text{not clipped}}$,
    $\dfrac{10,000}{n} = \dfrac{250}{10}$, $250n = 100,000$
    $n = \dfrac{100,000}{250} = 400$, assumes all hatchery fish
    survived and the hatchery fish mixed
    completely with native fish.

24. $\dfrac{x\ hr}{40\ miles} = \dfrac{1\ hr}{300\ miles}$
    $x = \dfrac{40}{300}$ hrs
    $x = \dfrac{2}{15}$ hrs
    $\dfrac{2}{15}$ hrs $\cdot \dfrac{60\ min}{1\ hr} = 8$ min
    The plane will reach the airport in 8 minutes.

25. a. 0.059(5,000,000,000)
       = \$295,000,000 annual interest.

    b. $\dfrac{295,000,000}{1\ yr} \cdot \dfrac{1\ yr}{365\ days} \cdot \dfrac{1\ day}{24\ hr}$
       $\cdot \dfrac{1\ hr}{60\ min} \cdot \dfrac{1\ min}{60\ sec}$
       $\approx$ \$9.35 per second

       Newspaper over-stated the interest by a
       small amount - or rounded to the nearest
       ten cents.

Exercises 5.4

1. a. 57 + 57 + 57 + 58 + 61 = 290
       290 ÷ 5 = 58, mean = 58 yrs
       median = 57 yrs, mode = 57 yrs

   b. 51 + 54 + 57 + 61 + 68 = 291
       291 ÷ 5 = 58.2, mean = 58.2 yrs
       median = 57 yrs, no mode

   c. 43 + 52 + 55 + 56 + 61 = 267
       267 ÷ 5 = 53.4 , mean = 53.4 yrs,
       median = 55 yrs, no mode

   d. 46 + 52 + 54 + 56 + 65 = 273
       273 ÷ 5 = 54.6, mean = 54.6 yrs
       median = 54 yrs, no mode

   Observation: Most presidents listed were in
   their 50s or 60s.

3. **a.** $345 + 350 + 350 + 365 + 375 + 385$

$+395 + 395 + 400 + 400 + 425 + 435$

$+450 + 460 = 5530$

$5530 \div 14 = 395$

Mean = \$395, Median = \$395, 3 Modes = \$350, \$395, and \$400.

   b. $425 + 425 + 430 + 435 + 450 + 475$

$+495 + 495 + 495 + 495 + 500$

$+525 + 539 + 550 = 6734$

$6734 \div 14 = \$481$

Mean = \$481, Median = \$495, Mode = \$495

Observation: Rents are about \$100 higher in the larger city.

5. To find the mean, add up all the values and divide by the number of values.

7. No; for example the mean of 5 and 7 is 6 and equals the mean of 4 and 8.

9. $A = \dfrac{1}{2}h(b + B)$

$= h \cdot \dfrac{(b + B)}{2}$

The area of a trapezoid is the product of the height and the average of the lengths of the parallel sides.

11. Is not; for example the median of 2, 3, 5 = 3 and the median of 0, 3, 25 = 3.

13. In both cases we are describing the middle of something.

15. The mode is the most frequently occurring data value(s).

17. One large piece of data tends to increase the value of the mean. However, there is little or no effect on the median or mode.

19. When we want to reduce the effect of large or small data.

21. If the mean is larger than the median, then there are some values that are comparatively very large.

23. Answers may vary. Possible answer:

Mean larger than median:
100, 126, 200
Have a relatively large number in the dataset.

Median larger than mean:
1, 100, 120
Have a relatively small number in the dataset.

25. **a.**   The mode for 1991 is 18 years and the mode for 2002 is 17 years.

   b.   The ages of unrestricted operation are becoming more uniform.

   c.   1991:

$(15(2) + 16(20) + 16.5 + 17(3)$

$+ 18(24) + 19) \div 51$

$= \dfrac{868.5}{51} \approx 17.03$

Mean ≈ 17.03 years

Median $= 17$ years

2002:

$(15(1) + 16(11) + 16.5(10) + 17(17)$

$+ 17.5(4) + 18(8) + 19(0)) \div 51$

$= \dfrac{859}{51} \approx 16.84$

Mean ≈ 16.84 years

Median $= 17$ years

   d.   The 2002 ages might be better because there are fewer younger drivers in several states.

27. GRCC = 4(5) + 3(3) + 2(1) = 31
    LCC = 5(5) + 0(3) + 4(1) = 29
    BCC = 1(5) + 7(3) + 4(1) = 30

29. $0.85 + 0.70 + 0.80 + 2(0.95) = 4.25$
    $4.25 \div 5 = 0.85$

31. $0.80 + 0.80 + 0.70 + 2(0.60) = 3.50$
    $3.50 \div 5 = 0.70$

33. $\dfrac{0.76 + 0.81 + 0.72 + 2(x)}{5} = 0.80$

$2.29 + 2x = 4, \quad 2x = 1.71, \quad x \approx 0.86$

35. $\dfrac{0.50+0.70+0.70+2(x)}{5}=0.80$

$1.9+2x=4,\quad 2x=2.1,\quad x=1.05$

Student needs over 100 on final, not possible.

37. a. $\left(\dfrac{0+5}{2},\ \dfrac{4+2}{2}\right)=\left(\dfrac{5}{2},\ \dfrac{6}{2}\right)=(2.5,\,3)$

b. $\left(\dfrac{-1+3}{2},\ \dfrac{3+(-3)}{2}\right)=\left(\dfrac{2}{2},\ \dfrac{0}{2}\right)$
$=(1,\,0)$

39. a. $\left(\dfrac{a+0}{2},\ \dfrac{0+b}{2}\right)=\left(\dfrac{a}{2},\ \dfrac{b}{2}\right)$

b. $\left(\dfrac{a+0}{2},\ \dfrac{a+0}{2}\right)=\left(\dfrac{a}{2},\ \dfrac{a}{2}\right)$

c. $\left(\dfrac{0+0}{2},\ \dfrac{b+0}{2}\right)=\left(0,\ \dfrac{b}{2}\right)$

41. Between $(0, 0)$ and $(0, 6)$ the midpoint is:
$\left(0,\ \dfrac{6}{2}\right)=(0,\ 3)$

Between $(0, 0)$ and $(9, 0)$ the midpoint is:
$\left(\dfrac{9}{2},\ 0\right)=(4.5,\ 0)$

Between $(0, 6)$ and $(9, 0)$ the midpoint is:
$\left(\dfrac{9}{2},\ \dfrac{6}{2}\right)=(4.5,\ 3)$

The centroid is at:
$\left(\dfrac{0+0+9}{3},\ \dfrac{0+6+0}{3}\right)=(3, 2)$

43. Between $(0, 0)$ and $(0, 6)$ the midpoint is:
$\left(0,\ \dfrac{6}{2}\right)=(0,\ 3)$

Between $(0, 0)$ and $(8, 0)$ the midpoint is:
$\left(\dfrac{8}{2},\ 0\right)=(4,\ 0)$

Between $(0, 6)$ and $(8, 0)$ the midpoint is:
$\left(\dfrac{8}{2},\ \dfrac{6}{2}\right)=(4,\ 3)$

The centroid is at:

$\left(\dfrac{0+0+8}{3},\ \dfrac{0+6+0}{3}\right)=(2\tfrac{2}{3},\ 2)$

Exercises 5.5

1. Value

3. Quantity

5. Value

7. Value

9. Value

11. a.

Quantity	Value	$Q\cdot V$
5 kg Peanuts	$8.80/kg	$44.00
2 kg Cashews	$24.20/kg	$48.40
7 kg	$13.20/kg	$92.40

b. The sum of the quantity column is the total kg of peanuts and cashews.

c. $92.40 \div 7$ kg = $13.20 per kg

d. The sum of the $Q\cdot V$ column is the total worth of peanuts and cashews.

13. a.

Quantity	Value	$Q\cdot V$
3 lbs Grapes	$0.98/lb	$2.94
5 lbs Potatoes	$0.49/lb	$2.45
2 lbs Broccoli	$0.89/lb	$1.78
10 lbs	$0.717/lb	$7.17

b. The sum of the quantity column is the total pounds of produce purchased.

c. $7.17 \div 10$ lbs $\approx 0.72 per pound

d. The sum of the $Q \cdot V$ column is the total amount of the purchase.

15. a.

Quantity	Value	$Q \cdot V$
15 dimes	$0.10	$1.50
20 quarters	$0.25	$5.00
35	$0.1857	$6.50

b. The sum of the quantity column is the total number of coins.

c. $6.50 \div 35 \approx 0.19

d. The sum of the $Q \cdot V$ column is the total dollar amount.

17. a.

Quantity	Value	$Q \cdot V$
$1500	0.09	$135
$1500	0.06	$90
$3000	0.075	$225

b. The sum of the quantity column is the total amount invested.

c. $225 \div $3000 = 0.075$, 7.5% average rate

d. The sum of the $Q \cdot V$ column is the total return on investment.

19. a.

Quantity	Value	$Q \cdot V$
100 lb	0.12	= 12.0
50 lb	0.15	= 7.5
150 lb	0.13	19.5

b. The sum of the quantity column is the total pounds of dog food.

c. $19.5 \div 150 = 0.13$, 13% protein

d. The sum of the $Q \cdot V$ column is the total pounds of protein.

21. a.

Quantity	Value	$Q \cdot V$
150 lb	0.10	15
25 lb	0	0
175 lb	0.086	15

b. The sum of the quantity column is the total pounds of dried grasses.

c. $15 \div 175 \approx 0.086$, 8.6% protein

d. The sum of the $Q \cdot V$ column is the total pounds of protein.

23. a.

Quantity	Value	$Q \cdot V$
5 hrs D	1	5
4 hrs C	2	8
3 hrs B	3	9
12 hrs	1.83	22

b. The sum of the quantity column is the total hours.

c. $22 \div 12$ hrs ≈ 1.83

d. The sum of the $Q \cdot V$ column is the total points.

25. a.

Quantity	Value	$Q \cdot V$
3 hr	80 kph	240 km
2 hr	30 kph	60 km
5 hr	60 kph	300 km

b. The sum of the quantity column is the total time.

c. $300 \text{ km} \div 5 \text{ hrs} = 60 \text{ kph}$

d.   The sum of the $Q \cdot V$ column is the total distance.

27. a.

Quantity	Value	$Q \cdot V$
150 mL	18	2700
100 mL	3	300
250 mL	12	3000

b.   The sum of the quantity column is total mL.

c.   The average molarity is
$3000 \div 250 \text{ ml} = 12$

d.   The sum of the $Q \cdot V$ column, divided by 1000, is the total moles of sulfuric acid. (The division is needed because molarity is in terms of liters, not milliliters.)

29.

Quantity	Value	$Q \cdot V$
A	0.05	0.05A
B or (15,000 - A)	0.08	0.08(15,000 - A) = 1200 - 0.08A
$15,000		1200 - 0.03A

a.   A = 0, B = $15,000
Earnings = $15,000(0.08) = $1200
Average rate = 0.08 = 8%

b.   A = $7,500, B = $7,500
Earnings = $7,500(0.05) + $7,500(0.08) =
$375 + $600
            = $975
Average rate = 975 ÷ 15,000
            = 0.065 = 6.5%

31.

Quantity	Value	$Q \cdot V$
200 Boeing	$30	$6000
x Nike	$46	$46x
200 + x		$19,800

$6000 + $46x = $19,800
46x = 13,800,   x = 300 shares of Nike.

33. $7200 - 200(40) = 7200 - 8000 = -800$

200 shares of Clorox costs $8000 which is more than the budgeted amount. No purchase of Maytag stock is possible.

35.

Quantity	Value	$Q \cdot V$
12	140°	1680
x	60°	60x
12 + x	103°	1680 + 60x

$103(12 + x) = 1680 + 60x$
$1236 + 103x = 1680 + 60x$
$43x = 444, \ x \approx 10.3 \text{ gal}$

37.

Quantity	Value	$Q \cdot V$
x	140°	140x
15	35°	525
15 + x	103°	525 + 140x

$103(15 + x) = 525 + 140x$
$1545 + 103x = 525 + 140x$
$1020 = 37x, \ x \approx 27.6 \text{ gal}$

39.

Quantity	Value	$Q \cdot V$
x	140°	140x
x	35°	35x
2x	t	175x

$2x(t) = 175x, \ t = \dfrac{175x}{2x}, \ t = 87.5°$

87.5° would be the result for any equal quantities because the quantity x drops out of the equation.

41.

Quantity	Value	$Q \cdot V$
20	1.85	37
x	3	3x
20 + x	2.00	37 + 3x

$2(20 + x) = 37 + 3x$
$40 + 2x = 37 + 3x$
$x = 3 \text{ hours}$

43.

Quantity	Value	Q·V
24	1.875	45
x	2.00	2x
24 + x	2.00	45 + 2x

$2(24 + x) = 45 + 2x$

$48 + 2x = 45 + 2x$

$3 = 0$ false

There are no real number solutions.

45.

Quantity	Value	Q·V
30 hrs	$5.80	$174.00
x hrs	$7.20	$7.20x
30 + x	$6.36	$174 + $7.20x

$6.36(30 + x) = $174 + $7.20x$

$190.80 + 6.36x = 174 + 7.20x$

$16.80 = 0.84x$, $x = 20$ hours

Neva must work 20 hours at the second job.

47. From exercise #45 with $7 average wage:

$7(30 + x) = $174 + $7.20x$

$210 + 7x = 174 + 7.20x$

$36 = 0.20x$, $x = 180$ hours, not reasonable since there are only 168 hours in a week.

49. 300 - x,

$x + (300 - x) = 300$

51. $15,000 - x,

$x + ($15,000 - x) = $15,000$

53. 16 - x,

$x + (16 - x) = 16$

55.

Quantity	Value	Q·V
300 - x Colombian	$8.35	(300 - x)$8.35
x Sumatran	$9.35	$9.35x
300	$8.75	$2625

$8.35(300 - x) + $9.35x = 2625

$2505 - 8.35x + 9.35x = 2625$

$x = 120$

300 - 120 = 180 lbs Colombian
120 lbs Sumatran

57.

Quantity	Value	Q·V
x peanuts	$10	$10x
50 - x cashews	$24	1200 - 24x
50		1200 - 14x

a. Place $12 as average value (below $24),
$50($12) = $1200 - $14x$,
solve for x.
$600 = 1200 - 14x$, $14x = 600$
$x \approx 42.9$ kg peanuts
$50 - 42.9 \approx 7.1$ kg cashews

b. Place $16 as average value (below $24),
$50($16) = $1200 - $14x$,
solve for x.
$800 = 1200 - 14x$, $14x = 400$
$x \approx 28.6$ kg peanuts
$50 - 28.6 \approx 21.4$ kg cashews

59.

Quantity	Value	Q·V
x @ 12%	0.12	0.12x
6000 − x @ 7%	0.07	0.07(6000 − x)
6000 @ 8%	0.08	480

$0.12x + 0.07(6000 - x) = 480$

$0.12x + 420 - 0.07x = 480$

$0.05x = 60$

$x = 1200$

The manager needs to mix 1200 pounds of the 12% protein with 4800 pounds of the 7% protein.

61. Using the table from exercise #29
The sum of Q·V equals the interest earned:

a. $1200 - 0.03A = $1060,
solve for A
$140 = 0.03A$, $A \approx 4666.67
$B \approx $15,000 - 4666.67
$\approx $10,333.33$

b. $1200 - 0.03A = $825
375 = 0.03A, A = $12,500
B = $15,000 - $12,500 = $2,500

63. In example 4 we multiply money invested by interest rate to find interest earned. In example 5 we multiply the number of shares by the price per share to get the money invested.

65. The average value is closer to the value of the item with the larger quantity.

67. If the quantity value is large compared to others it will have a larger effect.

69.

Quantity	Value	$Q \cdot V$
x ml	3	3x
300 ml	18	5400
x + 300	12	3x + 5400

$12(x + 300) = 3x + 5400$
$12x + 3600 = 3x + 5400$
$9x = 1800, \ x = 200$ ml

71. The greatest percent decrease is from $0.20 to $0.05. The smallest percent decrease was from $0.30 to $0.20.

$$\frac{0.50 - 1.00}{1.00} = -0.50, \ 50\% \text{ decrease}$$

$$\frac{0.30 - 0.50}{0.50} = -0.40, \ 40\% \text{ decrease}$$

$$\frac{0.20 - 0.30}{0.30} \approx -0.333, \ 33.3\% \text{ decrease}$$

$$\frac{0.05 - 0.20}{0.20} = 0.75, \ 75\% \text{ decrease}$$

Chapter 5 Review Exercises

1. $16x^2$ to $4x^4 = 4x^2(4)$ to $4x^2(x^2)$
$= 4$ to x^2

3. 5 feet = 60 inches
ratio is $60 : 18 = 6(10) : 6(3) = 10 : 3$

5. $\dfrac{\$42.49 - \$49.99}{\$49.99} = \dfrac{-7.50}{49.99} \approx -0.15$
15% decrease

7. $\dfrac{100 \text{ m}}{9 \text{ sec}} \cdot \dfrac{39.37 \text{ in}}{1 \text{ m}} \cdot \dfrac{1 \text{ ft}}{12 \text{ in}} \cdot \dfrac{1 \text{ mi}}{5280 \text{ ft}} \cdot \dfrac{3600 \text{ sec}}{1 \text{ hr}}$
$= \dfrac{14,173,200}{570,240} \text{ mi/hr}$
≈ 24.85 miles per hour

9. Answers will vary.

11. Let x = 1 share,
$5x + 5x + 1x = 121,000,$
$11x = 121,000,$
$x = 11,000$ cases per share

Pepsi = 5(11,000) = 55,000
Coca Cola = 5(11,000) = 55,000
Other = 1(11,000) = 11,000

13. $0.75x = 108, \ x = 144$

15. $0.30(1500) = 450$ cal,
$\dfrac{450 \text{ cal}}{1} \cdot \dfrac{1 \text{ g fat}}{9 \text{ cal}} = 50$ g

17. $\dfrac{1}{8} = \dfrac{30}{x}, \ x = 240$ in.
$240 \text{ in} \cdot \dfrac{1 \text{ ft}}{12 \text{ in}} = 20$ ft

19. $\dfrac{\text{total bats}}{\text{total tagged}} = \dfrac{\text{sample size}}{\text{tagged in sample}}$
$\dfrac{x}{250} = \dfrac{300}{15}$
$x = 250 \cdot \dfrac{300}{15}$
$x = 5000$
There are about 5000 bats in the cave.

21. $\dfrac{2}{3} = \dfrac{x}{17}$
$17 \cdot \dfrac{2}{3} = x$
$\dfrac{34}{3} = x$
$x = \dfrac{34}{3}$ or $x = 11\dfrac{1}{3}$

23.
$$\frac{2}{x} = 2,000,000$$

$$\frac{2}{2,000,000} = x$$

$$0.000001 = x$$

$$x = 0.000001$$

25.
$$\frac{2x+5}{7} = \frac{2x-1}{5}$$

$$5(2x+5) = 7(2x-1)$$

$$10x+25 = 14x-7$$

$$-4x = -32$$

$$x = 8$$

27.
$$\frac{V_1}{C_2} = \frac{V_2}{C_1}$$

$$C_1 \cdot \frac{V_1}{C_2} = V_2$$

$$V_2 = \frac{C_1 V_1}{C_2}$$

29.
$$\frac{x}{2} + \frac{1}{4} = 3 + \frac{x}{7}$$

$$\frac{x}{2} - \frac{x}{7} = 3 - \frac{1}{4}$$

$$\frac{7x}{14} - \frac{2x}{14} = \frac{12}{4} - \frac{1}{4}$$

$$\frac{5x}{14} = \frac{11}{4}$$

$$x = \frac{14}{5} \cdot \frac{11}{4}$$

$$x = \frac{77}{10} \quad \text{or} \quad x = 7.7$$

The equation is not a proportion because it is not in the form $\frac{a}{b} = \frac{c}{d}$.

31. The coordinates of point *A* are $(-3,4)$.

The x-coordinate of point *B* is 5. To find the y-coordinate, we can use similar triangles.

$$\frac{\text{base (large)}}{\text{height (large)}} = \frac{\text{base (small)}}{\text{height (small)}}$$

$$\frac{-6}{4} = \frac{2}{y}$$

$$-6y = 8$$

$$y = -\frac{8}{6}$$

$$y = -\frac{4}{3}$$

The coordinates of point *B* are $\left(5, -\frac{4}{3}\right)$ or $\left(5, -1\frac{1}{3}\right)$.

33. $\dfrac{4.2+4.3+4.3}{3} = \dfrac{12.8}{3} \approx 4.27$

The mean is about 4.27 grams.
The median is 4.3 grams.
The mode is 4.3 grams.

35. $\dfrac{6.8+6.8+6.8+6.9+6.9+7.0}{6}$

$$= \frac{41.2}{6} \approx 6.87$$

The mean is about 6.87 miles.
The median is 6.85 miles.
The mode is 6.8 miles.

37. a. $AD: \left(\dfrac{0+3}{2}, \dfrac{0+4}{2}\right) = \left(\dfrac{3}{2}, 2\right)$

$BC: \left(\dfrac{8+11}{2}, \dfrac{0+4}{2}\right) = \left(\dfrac{19}{2}, 2\right)$

$AB: \left(\dfrac{0+8}{2}, \dfrac{0+0}{2}\right) = (4,0)$

$CD: \left(\dfrac{11+3}{2}, \dfrac{4+4}{2}\right) = (7,4)$

$AC: \left(\dfrac{0+11}{2}, \dfrac{0+4}{2}\right) = \left(\dfrac{11}{2}, 2\right)$

$BD: \left(\dfrac{8+3}{2}, \dfrac{0+4}{2}\right) = \left(\dfrac{11}{2}, 2\right)$

b. $AD: m = \dfrac{4-0}{3-0} = \dfrac{4}{3}$

$AB: m = \dfrac{0-0}{8-0} = \dfrac{0}{8} = 0$

$$CD: \; m = \frac{4-4}{3-11} = \frac{0}{-8} = 0$$

$$BC: \; m = \frac{4-0}{11-8} = \frac{4}{3}$$

c. parallelogram

39. $x_c = \dfrac{-5+2+(-5)+2}{4} = \dfrac{-6}{4} = -\dfrac{3}{2}$

$y_c = \dfrac{5+5+(-1)+(-1)}{4} = \dfrac{8}{4} = 2$

The coordinates of the centroid are

$(x_c, y_c) = \left(-\dfrac{3}{2}, 2\right)$ or $(-1.5, 2)$.

41. a. Pair A: $0.4(5) + 1(1) = 2 + 1 = 3$

b. Pair B: $0.4(1) + 1(3) = 0.4 + 3 = 3.4$

c. Pair C: $0.4(2) + 1(2) = 0.8 + 2 = 2.8$

Pair C is the winner.

43.

Quantity	Value or Rate	Q · V
Number of coins	Value of coin, cents per coin	Cents
Time in hours	Miles per hour	Distance in miles
Dollars invested	Percent interest	Interest in dollars
Credit hours	Point value of grade per credit hour	Points
Number of cans	Cost per can	Cost
Number of pounds	Cost per pound	Cost
Hours worked	Wages per hour	Wages
Liters of acid	Molar value, moles per liter	Moles

45.

Quantity	Value	Q · V
4 hr	125 mph	500 mi
7.5 hr	200 mph	1500 mi
11.5 hr	≈174 mph	2000 mi

47.

Quanity	Value	Q · V
6 gal	88 octane	528 gal · octane
10 gal	92 octane	920 gal · octane
16 gal	90.5 octane	1448 gal · octane

49.

Quantity	Value	Q · V
x gal	130°	$130x$ gal · deg
9 gal	35°	315 gal · deg
$9+x$	103°	$103(9+x)$ gal · deg

$$130x + 315 = 103(9+x)$$
$$130x + 315 = 927 + 103x$$
$$27x = 612$$
$$x = \frac{68}{3} \text{ or } x = 22\frac{2}{3}$$

You would need to add $22\frac{2}{3}$ gallons of hot water.

51.

Quantity	Value	Q · V
x	0.08	$0.08x$
$25,000 - x$	0.18	$0.18(25,000 - x)$
25,000		2600

$$0.08x + 0.18(25,000 - x) = 2600$$
$$0.08x + 4500 - 0.18x = 2600$$
$$-0.1x = -1900$$
$$x = 19,000$$

Shawna borrowed $19,000 for school and has $6000 in credit card debt. Her average interest rate is $\dfrac{2600}{25000} = 0.104$ or 10.4%.

Chapter 5 Test

1. Answers will vary. For example: $\dfrac{3}{7}, \dfrac{9}{21}, \dfrac{12}{28}$

2. Answers will vary. For example: $\dfrac{3}{40}, \dfrac{15}{200}, \dfrac{30}{400}$

3. $\dfrac{7ab^3}{28a^2b^2} = \dfrac{b}{4a}$

4. $\dfrac{36(a+b)}{9} = \dfrac{9 \cdot 4(a+b)}{9}$

$\qquad = \dfrac{4(a+b)}{1}$

$\qquad = \dfrac{4a+4b}{1}$

5. $\dfrac{10 \text{ ft}}{24 \text{ in}} \cdot \dfrac{12 \text{ in}}{1 \text{ ft}} = \dfrac{5}{1}$

6. $1.25(x) = 105$, $x = 84$

7. $\dfrac{2 \text{ packs}}{1 \text{ day}} \cdot \dfrac{1 \text{ carton}}{20 \text{ packs}} \cdot \dfrac{\$35}{1 \text{ carton}} \cdot \dfrac{365 \text{ days}}{1 \text{ yr}}$

$= \$1{,}277.50$ per year

8. $\dfrac{3.5}{5} = \dfrac{x}{2}$, $x = \dfrac{7}{5}$, $x = 1.4$ ft

9. $\dfrac{2}{5} = \dfrac{13}{x}$, $2x = 65$, $x = 32.5$

10. $\dfrac{x+1}{4} = \dfrac{x-3}{3}$, $3(x+1) = 4(x-3)$

$3x+3 = 4x-12$, $x = 15$

11. $\dfrac{0.85 + 0.91 + x}{3} = 0.90$

$1.76 + x = 2.70$, $x = 0.94 = 94\%$

12. **a.** $\dfrac{12}{15} = \dfrac{x}{20}$, $x = \dfrac{240}{15}$, $x = 16$

b. $\dfrac{5}{9} = \dfrac{6}{x+6}$, $5(x+6) = 54$

$x + 6 = 10.8$, $x = 4.8$

13. $\$1.29 + \$1.29 + \$1.29 + \$1.29 + \$2.79 = \$7.95$

$\$7.95 \div 5 = \$1.59 = $ Mean

Median $= \$1.29$

14. $\$1.19 + \$1.29 + \$1.69 + \$1.69 + \$2.09 =$
$\$7.95$, $\$7.95 \div 5 = $ Mean $= \$1.59$

Median $= \$1.69$

15. Answers will vary

16. Midpoints:

$\left(\dfrac{3+7}{2}, \ \dfrac{2+8}{2} \right) = (5, \ 5)$

$\left(\dfrac{7+8}{2}, \ \dfrac{8+2}{2} \right) = (7.5, \ 5)$

$\left(\dfrac{8+3}{2}, \ \dfrac{2+2}{2} \right) = (5.5, \ 2)$

17. Centroid:

$\left(\dfrac{3+7+8}{3}, \ \dfrac{2+8+2}{3} \right) = (6, \ 4)$

18.

Quantity	Value	Q·V
$17,000	0.058	$986
$2,000	0.149	$298
$19,000 (Total debt)	$1284 ÷ $19,000 ≈ 0.068 (avg rate on total debt)	$1284 (Total interest on debt)

19.

Quantity	Value	Q·V
60 credits	3.40	204
x credits	4.00	4.00x
x + 60	3.50	4.00x + 204

$3.50(x + 60) = 4.00x + 204$
$3.50x + 210 = 4.00x + 204$
$6 = 0.50x$, $x = 12$ credits

$\dfrac{3.50 - 3.40}{3.40} = \dfrac{0.10}{3.40} \approx 0.029$

His GPA will have increased by about 2.9%.

20. The equation is not a proportion because it is

not in the form $\dfrac{a}{b} = \dfrac{c}{d}$.

$$\frac{8x}{3} - 1 = \frac{13}{3} + x$$

$$\frac{8x}{3} - x = \frac{13}{3} + 1$$

$$\frac{5x}{3} = \frac{16}{3}$$

$$x = \frac{3}{5} \cdot \frac{16}{3}$$

$$x = \frac{16}{5} \quad \text{or} \quad x = 3.2$$

21. The formula for percent change is $\dfrac{\text{change}}{\text{old}}$.

The percent change for revenue will be the same as the percent change for the number of columns.

$$\frac{\text{change}}{\text{old}} = \frac{10 - 9}{9} = \frac{1}{9}$$

The newspaper's revenue is increased by $\dfrac{1}{9}$.

$\dfrac{1}{9}$ is larger than $\dfrac{1}{10}$.

Chapter 6

Exercises 6.1

1. a. $-9 + 4 = -5$

   b. $-8 + 3 = -5$

   c. $-5 - 8 = -5 + (-8) = -13$

   d. $-7 - 3 = -7 + (-3) = -10$

   e. $-9 - (-5) = -9 + (+5) = -4$

   f. $3 - (-4) = 3 + (+4) = 7$

3. a. $8(-7) = -56$

   b. $-6(7) = -42$

   c. $-4(-9) = 36$

5. a. $-x^3 - 3x^2 + 5x + 5$

   b. $3x^3 + 2x^2 + 4x - 3$

7. a. $5a + 3b - 2c - 4a - 6c + 9b$

    $= 5a - 4a + 3b + 9b - 2c - 6c$

    $= a + 12b - 8c$

    trinomial

   b. $6m + 2n - 6p - 3m - 3n + 6p$

    $= 6m - 3m + 2n - 3n - 6p + 6p$

    $= 3m - n$

    binomial

   c. $8y + 5y + 5y - 5y + 8y = 21y$

    monomial

   d. $(x^2 + 3x) - (4x + 12)$

    $= x^2 + 3x - 4x - 12$

    $= x^2 - x - 12$

    trinomial

   e. $(x^2 - 2x + 3) - (2x^2 - 4x + 6)$

    $= x^2 - 2x + 3 - 2x^2 + 4x - 6$

    $= x^2 - 2x^2 - 2x + 4x + 3 - 6$

    $= -x^2 + 2x - 3$

    trinomial

9. a. $x^2 + 2x + 3x + 6 = x^2 + 5x + 6$

    trinomial

   b. $3x^2 + 6x + x + 2 = 3x^2 + 7x + 2$

    trinomial

   c. $5 - 4(x - 3) = 5 - 4x + 12$

                 $= -4x + 17$

    binomial

   d. $6x^2 + 2x + 3x + 1 = 6x^2 + 5x + 1$

    trinomial

   e. $a^2 - ab + ab - b^2 = a^2 - b^2$

    binomial

11. Length: $a + b$

    Width: $b + b = 2b$

    Perimeter: $2(a + b) + 2(2b)$

               $= 2a + 2b + 4b$

               $= 2a + 6b$

    Area: $(2b)(a + b) = 2ab + 2b^2$

13. Length: $x + 1 + 1 = x + 2$

    Width: $x + 1$

    Perimeter: $2(x + 2) + 2(x + 1)$

               $= 2x + 4 + 2x + 2$

               $= 4x + 6$

    Area: $(x + 2)(x + 1)$

           $= x^2 + x + 2x + 2$

           $= x^2 + 3x + 2$

15. a. $2a(a + 2b) = 2a^2 + 4ab$

    b. $2b(2a + b) = 4ab + 2b^2$

    c. $x(x + 3) = x^2 + 3x$

    d. $x(2x + 1) = 2x^2 + x$

17. $4x^2y^4 \cdot x^2 = 4x^{2+2}y^4 = 4x^4y^4$

$4x^2y^4 \cdot (-3xy) = -12x^{2+1}y^{4+1} = -12x^3y^5$

$4x^2y^4 \cdot (-4y^2) = -16x^2y^{4+2} = -16x^2y^6$

Multiply	x^2	$-3xy$	$-4y^2$
$4x^2y^4$	$4x^4y^4$	$-12x^3y^5$	$-16x^2y^6$

19. $-5x^3y^4 \cdot x^2 = -5x^{3+2}y^4 = -5x^5y^4$

$-5x^3y^4 \cdot (-2xy) = +10x^{3+1}y^{4+1}$

$\qquad\qquad = +10x^4y^5$

$-5x^3y^4 \cdot (-y^2) = +5x^3y^{4+2} = +5x^3y^6$

Multiply	x^2	$-2xy$	$-y^2$
$-5x^3y^4$	$-5x^5y^4$	$+10x^4y^5$	$+5x^3y^6$

21. a. $2x(x^2+3x) = 2x^3+6x^2$

    b. $x^2(x-1) = x^3-x^2$

    c. $x^2(x^2+2x+1) = x^4+2x^3+x^2$

    d. $ab(b^2-1) = ab^3-ab$

    e. $b^2(a-b) = ab^2-b^3$

    f. $a^2(1+b-b^2) = a^2+a^2b-a^2b^2$

23. a. $5x-3x(1-x) = 5x-3x+3x^2$

$\qquad\qquad\qquad = 2x+3x^2$

$\qquad\qquad\qquad = 3x^2+2x$

    b. $4a+2a(3-a) = 4a+6a-2a^2$

$\qquad\qquad\qquad = 10a-2a^2$

$\qquad\qquad\qquad = -2a^2+10a$

25. a. $4b-2b(5-b) = 4b-10b+2b^2$

$\qquad\qquad\qquad = -6b+2b^2$

$\qquad\qquad\qquad = 2b^2-6b$

    b. $6x-x(x+1) = 6x-x^2-x$

$\qquad\qquad\qquad = -x^2+5x$

27. a. $2y(-x+2y)+x(x-2y)$

$\qquad = -2xy+4y^2+x^2-2xy$

$\qquad = x^2-2xy-2xy+4y^2$

$\qquad = x^2-4xy+4y^2$

    b. $x^2-4y^2-2xy+2xy = x^2-4y^2$

    c. $x^3+2x^2+x+x^2+2x+1$

$\qquad = x^3+2x^2+x^2+x+2x+1$

$\qquad = x^3+3x^2+3x+1$

    d. $x(4+4x+x^2)-2(4+4x+x^2)$

$\qquad = 4x+4x^2+x^3-8-8x-2x^2$

$\qquad = x^3+4x^2-2x^2+4x-8x-8$

$\qquad = x^3+2x^2-4x-8$

    e. $b^3-ab^2+a^2b+a^3-a^2b+ab^2$

$\qquad = a^3+a^2b-a^2b-ab^2+ab^2+b^3$

$\qquad = a^3+b^3$

29.

Factor	y^2	$+2xy$	$+x^2$
y	y^3	$+2xy^2$	$+x^2y$

31.

Factor	a	$-2b$	$+3b^2$
$2ab$	$2a^2b$	$-4ab^2$	$+6ab^3$

33. gcf $= 2a$

Factor	1	$-2b$
$2a$	$2a$	$-4ab$

35. gcf $= ab^2$

Factor	1	$-a$	$-ab$
ab^2	ab^2	$-a^2b^2$	$-a^2b^3$

37. gcf $= x$

$\qquad x^3+4x^2+4x = x(x^2+4x+4)$

39. gcf $= b$

$\qquad a^2b+ab^2+b^3 = b(a^2+ab+b^2)$

41. gcf $= 2x$

$\qquad 6x^2+2x = 2x(3x+1)$

43. gcf $= 3y$

$$15y^2 - 3y = 3y(5y - 1)$$

45. gcf $= 5xy$

$$15x^2 y + 10xy^2 = 5xy(3x + 2y)$$

47. $-4x - 12 = 4(-x - 3)$

$$-4x - 12 = -4(x + 3)$$

49. $-2xy + 4y^2 = 2y(-x + 2y)$

$$-2xy + 4y^2 = -2y(x - 2y)$$

51. $-12x^2 - 8x - 8 = 4(-3x^2 - 2x - 2)$

$$-12x^2 - 8x - 8 = -4(3x^2 + 2x + 2)$$

53. $-y^2 + 4y^3 - 8y^4 = y^2(-1 + 4y - 8y^2)$

$$-y^2 + 4y^3 - 8y^4 = -y^2(1 - 4y + 8y^2)$$

55. 2 terms

57. 1 term

59. 3 terms

61. 3 factors

63. 4 factors

65. 4 factors

67. The two expressions are equal because of the distributive property.

69. The distributive property, $a(b + c) = ab + ac$, changes the product of two <u>factors</u> into the <u>sum</u> of two <u>terms</u>.

71. Two terms are like terms if they have the same variables and the exponents on corresponding variables are the same in each term.

73. The greatest common factor is the product of all the factors that are common to each term.

75. **a.** $111 = 3 \cdot 37$

  b. $91 = 7 \cdot 13$

77. **a.** $36 = 2 \cdot 2 \cdot 3 \cdot 3$

  $99 = 3 \cdot 3 \cdot 11$

  gcf $= 3 \cdot 3 = 9$

$$\frac{36}{99} = \frac{9 \cdot 4}{9 \cdot 11} = \frac{4}{11}$$

 b. $66 = 2 \cdot 3 \cdot 11$

  $990 = 2 \cdot 3 \cdot 3 \cdot 5 \cdot 11$

  gcf $= 2 \cdot 3 \cdot 11 = 66$

$$\frac{66}{990} = \frac{66 \cdot 1}{66 \cdot 15} = \frac{1}{15}$$

 c. $185 = 5 \cdot 37$

  $999 = 3 \cdot 3 \cdot 3 \cdot 37$

  gcf $= 37$

$$\frac{185}{999} = \frac{5 \cdot 37}{27 \cdot 37} = \frac{5}{27}$$

 d. gcf $= m$

$$\frac{mn}{mp} = \frac{m \cdot n}{m \cdot p} = \frac{n}{p}$$

 e. $4np = 2 \cdot 2 \cdot n \cdot p$

  $24mn = 2 \cdot 2 \cdot 2 \cdot 3 \cdot m \cdot n$

  gcf $= 2 \cdot 2 \cdot n = 4n$

$$\frac{4np}{24mn} = \frac{4n \cdot p}{4n \cdot 6m} = \frac{p}{6m}$$

79. **a.** $P = 1.75a + 0.75a + 1.75a + 0.75a$

  $= (1.75 + 0.75 + 1.75 + 0.75)a$

  $= 5a$

 b. $P = x + x + x + x = 4x$

 c. $P = (2x - 3) + x + 2x$

  $= 2x - 3 + x + 2x$

  $= 5x - 3$

 d. $P = 0.5\pi + 1.5 + 0.5\pi + 1.5$

  $= (1.5 + 1.5) + (0.5\pi + 0.5\pi)$

  $= 3 + (0.5 + 0.5)\pi$

  $= 3 + \pi$

81. **a.** $P = 10a + 6c$

  $= 2(5a + 3c)$

  The width is $3c$.

$$A = (5a)(3c) = 15ac$$

b. $P = 20y = 4(5y)$

Each side of the square is $5y$.

$$A = (5y)(5y) = 25y^2$$

c. $P = 3x + 5y$

$$= 2\left(\frac{3}{2}x + \frac{5}{2}y\right)$$

$$= 2(1.5x + 2.5y)$$

The length is $2.5y$.

$$A = (1.5x)(2.5y) = 3.75xy$$

Exercises 6.2

1. $(x+1)(x+1) = x^2 + 2x + 1$

3. $(2a+b)(a+2b)$

$\quad = 2a^2 + ab + 4ab + 2b^2$

$\quad = 2a^2 + 5ab + 2b^2$

5. $(a+2b)(a+2b)$

$\quad = a^2 + 2ab + 2ab + 4b^2$

$\quad = a^2 + 4ab + 4b^2$

7.

Multiply	$2x$	$+5$
x	$2x^2$	$+5x$
-4	$-8x$	-20

$(x-4)(2x+5) = 2x^2 + 5x - 8x - 20$

$\qquad\qquad\qquad = 2x^2 - 3x - 20$

9.

Multiply	x	-2
$2x$	$2x^2$	$-4x$
$+3$	$+3x$	-6

$(2x+3)(x-2) = 2x^2 + 3x - 4x - 6$

$\qquad\qquad\qquad = 2x^2 - x - 6$

11.

Multiply	x	-2
$2x$	$2x^2$	$-4x$
-1	$-1x$	$+2$

$(2x-1)(x-2) = 2x^2 - x - 4x + 2$

$\qquad\qquad\qquad = 2x^2 - 5x + 2$

13.

Multiply	x	-4
$5x$	$5x^2$	$-20x$
-1	$-1x$	$+4$

$(5x-1)(x-4) = 5x^2 - 1x - 20x + 4$

$\qquad\qquad\qquad = 5x^2 - 21x + 4$

15. a. $(x-2)(x-2) = x^2 - 2x - 2x + 4$

$\qquad\qquad\qquad = x^2 - 4x + 4$

b. $(x+2)(x-2) = x^2 + 2x - 2x - 4$

$\qquad\qquad\qquad = x^2 - 4$

17. a. $(a+5)(a+5) = a^2 + 5a + 5a + 25$

$\qquad\qquad\qquad = a^2 + 10a + 25$

b. $(b+5)(b-5) = b^2 + 5b - 5b - 25$

$\qquad\qquad\qquad = b^2 - 25$

19. a. $(a+b)(a-b) = a^2 + ab - ab - b^2$

$\qquad\qquad\qquad = a^2 - b^2$

b. $(a-b)(a-b) = a^2 - ab - ab + b^2$

$\qquad\qquad\qquad = a^2 - 2ab + b^2$

21. a. $(x+1)(x+7) = x^2 + x + 7x + 7$

$\qquad\qquad\qquad = x^2 + 8x + 7$

b. $(x+1)(x-7) = x^2 + x - 7x - 7$

$\qquad\qquad\qquad = x^2 - 6x - 7$

23. a. $(b+7)(b+7) = b^2 + 7b + 7b + 49$

$\qquad\qquad\qquad = b^2 + 14b + 49$

b. $(a+7)(a-7) = a^2 + 7a - 7a - 49$

$\qquad\qquad\qquad = a^2 - 49$

25. a. $(x+y)(x+y) = x^2 + xy + xy + y^2$

$\qquad\qquad\qquad = x^2 + 2xy + y^2$

b. $(x-y)(x-y) = x^2 - xy - xy + y^2$

$\qquad\qquad\qquad = x^2 - 2xy + y^2$

27. $15a, 16a, 17a, 18b, 19b,$
$20b, 23a, 24b, 25a, 25b$

29. pst;
$$(2x+3)(2x+3) = 4x^2 + 6x + 6x + 9$$
$$= 4x^2 + 12x + 9$$

31. ds;
$$(2x-3)(2x+3) = 4x^2 - 6x + 6x - 9$$
$$= 4x^2 - 9$$

33. neither;
$$(2x-3)(3-2x) = 6x - 9 - 4x^2 + 6x$$
$$= -4x^2 + 12x - 9$$

35. pst;
$$(3x-2)(3x-2) = 9x^2 - 6x - 6x + 4$$
$$= 9x^2 - 12x + 4$$

37. ds;
$$(3x+2)(3x-2) = 9x^2 + 6x - 6x - 4$$
$$= 9x^2 - 4$$

39. pst;
$$(x+5)^2 = (x+5)(x+5)$$
$$= x^2 + 5x + 5x + 25$$
$$= x^2 + 10x + 25$$

41. pst;
$$(a-6)^2 = (a-6)(a-6)$$
$$= a^2 - 6a - 6a + 36$$
$$= a^2 - 12a + 36$$

43. $(x-1)(x+12) = x^2 + 11x - 12$

45. $(x-3)(x+4) = x^2 + 1x - 12$

47. $(x+2)(x-6) = x^2 - 4x - 12$

49. $(x+1)(x+12)$; $(x-1)(x-12)$;
$(x+2)(x+6)$; $(x-2)(x-6)$;
$(x+3)(x+4)$; $(x-3)(x-4)$

51. $(x+1)(x+20)$; $(x-1)(x-20)$;
$(x+2)(x+10)$; $(x-2)(x-10)$;
$(x+4)(x+5)$; $(x-4)(x-5)$

53. The value of n will be the sum of 2 numbers that multiply to +24. Possible values for n are: $1 + 24 = 25$; $2 + 12 = 14$; $3 + 8 = 11$; $4 + 6 = 10$.

55. a. $(x+1)(x+8) = x^2 + x + 8x + 8$
$$= x^2 + 9x + 8$$

   b. $(x+1)(x-8) = x^2 + x - 8x - 8$
$$= x^2 - 7x - 8$$

57. a. $(x+2)(x+4) = x^2 + 2x + 4x + 8$
$$= x^2 + 6x + 8$$

   b. $(x+2)(x-4) = x^2 + 2x - 4x - 8$
$$= x^2 - 2x - 8$$

59. a. $(2x-3)(3x-2) = 6x^2 - 9x - 4x + 6$
$$= 6x^2 - 13x + 6$$

   b. $(2x+3)(3x-2) = 6x^2 + 9x - 4x - 6$
$$= 6x^2 + 5x - 6$$

61. a. $(6x+1)(x+6) = 6x^2 + x + 36x + 6$
$$= 6x^2 + 37x + 6$$

   b. $(6x-1)(x+6) = 6x^2 - x + 36x - 6$
$$= 6x^2 + 35x - 6$$

63. a. $(2x+5)(x+1) = 2x^2 + 5x + 2x + 5$
$$= 2x^2 + 7x + 5$$

   b. $(2x+5)(x-1) = 2x^2 + 5x - 2x - 5$
$$= 2x^2 + 3x - 5$$

65. a. $(2x-1)(x+5) = 2x^2 - x + 10x - 5$
$$= 2x^2 + 9x - 5$$

   b. $(2x+1)(x+5) = 2x^2 + x + 10x + 5$
$$= 2x^2 + 11x + 5$$

67. Written response. Answers may vary.

69. Written response. Answers may vary.

71. The coefficient -2 on the middle term is obtained by adding the results of the products $3 \cdot x$ and $-5 \cdot x$.

73. $2(x+3)^2 = 2\left(x^2+3x+3x+9\right)$

$\qquad = 2\left(x^2+6x+9\right)$

$\qquad = 2x^2+12x+18$

75. $3(x-5)^2 = 3\left(x^2-5x-5x+25\right)$

$\qquad = 3\left(x^2-10x+25\right)$

$\qquad = 3x^2-30x+75$

77. $4(5-x)^2 = 4\left(25-5x-5x+x^2\right)$

$\qquad = 4\left(x^2-10x+25\right)$

$\qquad = 4x^2-40x+100$

79. $5(3-2x)^2 = 5\left(9-6x-6x+4x^2\right)$

$\qquad = 5\left(4x^2-12x+9\right)$

$\qquad = 20x^2-60x+45$

81. Exponents cannot be distributed across a sum or a difference.

$(x-a)^2 = (x-a)(x-a)$

$\qquad = x^2-ax-ax+a^2$

$\qquad = x^2-2ax+a^2$

83. a.

Multiply	x^2	$+2x$	$+4$
x	x^3	$+2x^2$	$+4x$
-2	$-2x^2$	$-4x$	-8

$(x-2)\left(x^2+2x+4\right)$

$= x^3+2x^2+4x-2x^2-4x-8$

$= x^3-8$

b.

Multiply	x^2	$-2x$	$+1$
x	x^3	$-2x^2$	$+1x$
-1	$-x^2$	$+2x$	-1

$(x-1)\left(x^2-2x+1\right)$

$= x^3-2x^2+1x-x^2+2x-1$

$= x^3-3x^2+3x-1$

c.

Multiply	a^2	$+ab$	$+b^2$
a	a^3	$+a^2b$	$+ab^2$
$-b$	$-a^2b$	$-ab^2$	$-b^3$

$(a-b)\left(a^2+ab+b^2\right)$

$= a^3+a^2b+ab^2-a^2b-ab^2-b^3$

$= a^3-b^3$

d.

Multiply	x^2	$-xy$	$+y^2$
x	x^3	$-x^2y$	$+xy^2$
$+y$	$+x^2y$	$-xy^2$	$+y^3$

$(x+y)\left(x^2-xy+y^2\right)$

$= x^3-x^2y+xy^2+x^2y-xy^2+y^3$

$= x^3+y^3$

e.

Multiply	x^2	$+2xy$	$+y^2$
x	x^3	$+2x^2y$	$+xy^2$
$+y$	$+x^2y$	$+2xy^2$	$+y^3$

$(x+y)\left(x^2+2xy+y^2\right)$

$= x^3+2x^2y+xy^2+x^2y+2xy^2+y^3$

$= x^3+3x^2y+3xy^2+y^3$

f.

Multiply	a^2	$+2ab$	$+b^2$
a	a^3	$+2a^2b$	$+ab^2$
$+b$	$+a^2b$	$+2ab^2$	$+b^3$

$(a+b)\left(a^2+2ab+b^2\right)$

$= a^3+2a^2b+ab^2+a^2b+2ab^2+b^3$

$= a^3+3a^2b+3ab^2+b^3$

g. In the table, like terms are located on diagonals.

Exercises 6.3

1. $(2x+1)(x+2) = 2x^2+x+4x+2$

$\qquad\qquad\qquad = 2x^2+5x+2$

3. $(2x+1)(2x+1) = 4x^2 + 2x + 2x + 1$
$$= 4x^2 + 4x + 1$$

5.

Factor	x	$+5$
x	x^2	$+5x$
$+4$	$+4x$	$+20$

$(x+4)(x+5) = x^2 + 5x + 4x + 20$
$$= x^2 + 9x + 20$$

7.

Factor	x	$+2$
x	x^2	$+2x$
-10	$-10x$	-20

$(x-10)(x+2) = x^2 + 2x - 10x - 20$
$$= x^2 - 8x - 20$$

9.

Factor	$6x$	$+1$
x	$6x^2$	$+x$
-3	$-18x$	-3

$(x-3)(6x+1) = 6x^2 + x - 18x - 3$
$$= 6x^2 - 17x - 3$$

11. The middle term is obtained by adding the products $3 \cdot x$ and $-5 \cdot x$.
$3x + (-5x) = -2x$

13. $x^2 + 8x + 12 = (x+6)(x+2)$

15. $x^2 - 13x + 12 = (x-12)(x-1)$

17. $x^2 + 4x - 12 = (x+6)(x-2)$

19. $x^2 + 7x + 12 = (x+3)(x+4)$

21. $x^2 + x - 12 = (x+4)(x-3)$

23. $x^2 - 11x - 12 = (x-12)(x+1)$

25. perfect square trinomial;
$x^2 + 6x + 9 = (x+3)^2$

27. $x^2 + 11x + 30 = x^2 + 5x + 6x + (5)(6)$
$$= (x+5)(x+6)$$

29. $x^2 + 13x - 30 = x^2 + 15x - 2x + (15)(-2)$
$$= (x+15)(x-2)$$

31. $x^2 - 6x - 16 = x^2 - 8x + 2x + (-8)(2)$
$$= (x-8)(x+2)$$

33. $x^2 + 15x - 16 = x^2 + 16x - x + (15)(-1)$
$$= (x+15)(x-1)$$

35. difference of two squares;
$x^2 + 0x - 25 = (x-5)(x+5)$

37. Diagonal product $= 24x^2$,
sum $= 11x$
Middle terms are $3x$ and $8x$.

Factor	$2x$	$+3$
x	$2x^2$	$+3x$
$+4$	$+8x$	$+12$

$2x^2 + 11x + 12 = (2x+3)(x+4)$

39. Diagonal product $= -18x^2$,
sum $= -3x$
Middle terms are $3x$ and $-6x$.

Factor	$2x$	$+3$
x	$2x^2$	$+3x$
-3	$-6x$	-9

$2x^2 - 3x - 9 = (x-3)(2x+3)$

41. Diagonal product $= -6n^2$,
sum $= n$
Middle terms are $3n$ and $-2n$.

Factor	$2n$	$+3$
n	$2n^2$	$+3n$
-1	$-2n$	-3

$2n^2 + n - 3 = (n-1)(2n+3)$

43. Diagonal product = $-6x^2$,
sum = $5x$
Middle terms are $6x$ and $-x$.

Factor	$3x$	-1
x	$3x^2$	$-x$
$+2$	$+6x$	-2

$3x^2 + 5x - 2 = (x+2)(3x-1)$

45. Diagonal product = $-12a^2$,
sum = $-11a$
Middle terms are $-12a$ and a.

Factor	$3a$	$+1$
a	$3a^2$	$+a$
-4	$-12a$	-4

$3a^2 - 11a - 4 = (a-4)(3a+1)$

47. difference of two squares;
$9x^2 + 0x - 49 = (3x-7)(3x+7)$

49. difference of two squares;
$16x^2 + 0x - 9 = (4x-3)(4x+3)$

51. Diagonal product = $-12x^2$,
sum = x
Middle terms are $4x$ and $-3x$.

Factor	$2x$	-1
$3x$	$6x^2$	$-3x$
$+2$	$+4x$	-2

$6x^2 + x - 2 = (3x+2)(2x-1)$

53. Diagonal product = $-36x^2$,
sum = $5x$
Middle terms are $9x$ and $-4x$.

Factor	$2x$	$+3$
$3x$	$6x^2$	$+9x$
-2	$-4x$	-6

$6x^2 + 5x - 6 = (3x-2)(2x+3)$

55. Diagonal product = $-10n^2$,
sum = $9n$
Middle terms are $10n$ and $-n$.

Factor	n	$+5$
$2n$	$2n^2$	$+10n$
-1	$-n$	-5

$2n^2 + 9n - 5 = (2n-1)(n+5)$

Mid-Chapter 6 Test

1. a. $7b - 8c + 3a + 4b - 5c - 6a$
$= 3a - 6a + 7b + 4b - 8c - 5c$
$= -3a + 11b - 13c$
trinomial

b. $x^2 - 4x + x^2 - 6$
$= x^2 + x^2 - 4x - 6$
$= 2x^2 - 4x - 6$
trinomial

c. $4xy^2 + x^3 y^2 - 3x^2 y + x^3 y^2 - 2xy^2$
$= x^3 y^2 + x^3 y^2 - 3x^2 y + 4xy^2 - 2xy^2$
$= 2x^3 y^2 - 3x^2 y + 2xy^2$
trinomial

d. $(5a - 3b - 2c) - (3a + 4b - 6c)$
$= 5a - 3b - 2c - 3a - 4b + 6c$
$= 5a - 3a - 3b - 4b - 2c + 6c$
$= 2a - 7b + 4c$
trinomial

e. $x(x^2 + 5x + 25) - 5(x^2 + 5x + 25)$
$= x^3 + 5x^2 + 25x - 5x^2 - 25x - 125$
$= x^3 - 125$
binomial

f. $9 - 3x(x+3) = 9 - 3x^2 - 9x$
$\qquad\qquad\qquad = -3x^2 - 9x + 9$
trinomial

g. $8 - 4(4-x) = 8 - 16 + 4x$
$\qquad\qquad\quad = 4x - 8$
binomial

2. a. $\text{gcf} = 2$

$$6x^2 - 2x + 8 = 2\left(3x^2 - x + 4\right)$$

b. $\text{gcf} = a$

$$2abc - 3ac + 4ab = a\left(2bc - 3c + 4b\right)$$

3. Terms are added or subtracted, such as $a + b$ or $a - b$. Factors are multiplied, such as $a \cdot b$.

4. a. Length $= 2a + b$

Width $= a + 2b$

Perimeter $= 2(2a + b) + 2(a + 2b)$

$\qquad = 4a + 2b + 2a + 4b$

$\qquad = 6a + 6b$

Area $= (2a + b)(a + 2b)$

$\qquad = 2a^2 + 4ab + ab + 2b^2$

$\qquad = 2a^2 + 5ab + 2b^2$

b. Length $= x + 2$

Width $= x + 1$

Perimeter $= 2(x + 2) + 2(x + 1)$

$\qquad = 2x + 4 + 2x + 2$

$\qquad = 4x + 6$

Area $= (x + 2)(x + 1)$

$\qquad = x^2 + x + 2x + 2$

$\qquad = x^2 + 3x + 2$

5. The tiles can be arranged as:

$$(a + 2b)(3a + 2b)$$

6.

Multiply	$2x$	-5
$3x$	$6x^2$	$-15x$
$+2$	$+4x$	-10

$(3x + 2)(2x - 5) = 3x^2 - 15x + 4x - 10$

$\qquad\qquad\qquad\quad = 3x^2 - 11x - 10$

7. The diagonal product is $\left(6x^2\right)(-10) = -60x^2$

8.

Multiply	$3x^2$	$-2x$	$+1$
x	$3x^3$	$-2x^2$	$+1x$
-2	$-6x^2$	$+4x$	-2

$(x - 2)(3x^2 - 2x + 1)$

$= 3x^3 - 2x^2 + x - 6x^2 + 4x - 2$

$= 3x^3 - 8x^2 + 5x - 2$

9. The like terms are:

$-6x^2$ and $-2x^2$;

$4x$ and x

10. $(x + 3)(x + 5) = x^2 + 5x + 3x + 15$

$\qquad\qquad\qquad\quad = x^2 + 8x + 15$

11. $(x - 4)(3x + 5) = 3x^2 + 5x - 12x - 20$

$\qquad\qquad\qquad\qquad = 3x^2 - 7x - 20$

12. $(2x + 7)(3x - 1) = 6x^2 - 2x + 21x - 7$

$\qquad\qquad\qquad\qquad = 6x^2 + 19x - 7$

13. $(x - 3)(x + 3) = x^2 + 3x - 3x - 9$

$\qquad\qquad\qquad\quad = x^2 - 9$

14. $(2x - 5)^2 = (2x - 5)(2x - 5)$

$\qquad\qquad\quad = 4x^2 - 10x - 10x + 25$

$\qquad\qquad\quad = 4x^2 - 20x + 25$

15. a. neither

b. neither

c. ds

d. pst

16. $(x \pm 1)(x \pm 10) = x^2 \pm 11x + 10$,

$(x \mp 1)(x \pm 10) = x^2 \pm 9x - 10$,

$(x \pm 2)(x \pm 5) = x^2 \pm 7x + 10$,

$(x \mp 2)(x \pm 5) = x^2 \pm 3x - 10$

17. (a) $2a^2 + 5ab + 2b^2 = (a + 2b)(2a + b)$

(b) $x^2 + 3x + 2 = (x + 1)(x + 2)$

18. $3a^2 + 8ab + 4b^2 = (a + 2b)(3a + 2b)$

19. a.

Factor	$3x$	-4
$2x$	$6x^2$	$-8x$
$+5$	$+15x$	-20

$6x^2 + 7x - 20 = (2x+5)(3x-4)$

b.

Factor	x	-4
$6x$	$6x^2$	$-24x$
$+5$	$+5x$	-20

$6x^2 - 19x - 20 = (6x+5)(x-4)$

20. a. Diagonal product $= 35x^2$,
sum $= 12x$
Middle terms are $7x$ and $5x$.

Factor	x	$+7$
x	x^2	$+7x$
$+5$	$+5x$	$+35$

$x^2 + 12x + 35 = (x+7)(x+5)$

b. Diagonal product $= -14x^2$,
sum $= -5x$
Middle terms are $-7x$ and $2x$.

Factor	x	$+2$
x	x^2	$+2x$
-7	$-7x$	-14

$x^2 - 5x - 14 = (x-7)(x+2)$

c. Diagonal product $= -60x^2$,
sum $= -17x$
Middle terms are $-20x$ and $3x$.

Factor	$3x$	-10
$2x$	$6x^2$	$-20x$
$+1$	$+3x$	-10

$6x^2 - 17x - 10 = (2x+1)(3x-10)$

Exercises 6.4

1. $x^2 - 4 = (x-2)(x+2)$
difference of two squares

3. $4x^2 - 16 = 4(x^2 - 4)$
$\qquad = 4(x-2)(x+2)$
difference of two squares

5. $x^2 + 12x + 36 = (x+6)^2$
perfect square trinomial

7. $4x^2 + 8x + 4 = 4(x^2 + 2x + 1)$
$\qquad = 4(x+1)^2$
perfect square trinomial

9. $x^2 + 4$
sum of two squares – cannot be factored.

11. $x^2 - 6x + 9 = (x-3)^2$
perfect square trinomial

13. $3x^2 + 12x + 9 = 3(x^2 + 4x + 3)$

Diagonal product $= 3x^2$,
sum $= 4x$
Middle terms are $3x$ and x.

Factor	x	$+3$
x	x^2	$+3x$
$+1$	$+x$	$+3$

$3x^2 + 12x + 9 = 3(x+3)(x+1)$

15. $3x^2 - 27 = 3(x^2 - 9)$
$\qquad = 3(x-3)(x+3)$

17. $3a^2 + 5a - 4$
Diagonal product $= -12a^2$,
sum $= 5a$
cannot be factored

19. $25x^2 - 36 = (5x-6)(5x+6)$

21. $3x^2 + 10x - 5$
Diagonal product $= -15x^2$,
sum $= 10x$
cannot be factored

23. $5x^2 - 10x + 5 = 5(x^2 - 2x + 1)$
$\qquad = 5(x-1)^2$

25. $18x^2 - 50 = 2(9x^2 - 25)$
$\qquad = 2(3x-5)(3x+5)$

27. $3x^2 - 30x + 75 = 3(x^2 - 10x + 25)$
$$= 3(x - 5)^2$$

29. $3x^2 + 6x + 12 = 3(x^2 + 2x + 4)$

Diagonal product $= 4x^2$, sum $= 2x$
cannot be factored further

31. $8x^2 - 5x - 1$

Diagonal product $= -8x^2$, sum $= -5x$
cannot be factored

33. $6x^2 - 4x - 3$

Diagonal product $= -18x^2$, sum $= -4x$
cannot be factored

35. $2x^3 + x^2 - 3x = x(2x^2 + x - 3)$

Diagonal product $= -6x^2$, sum $= x$
Middle terms are $3x$ and $-2x$.

Factor	x	-1
$2x$	$2x^2$	$-2x$
$+3$	$+3x$	-3

$2x^3 + x^2 - 3x = x(2x + 3)(x - 1)$

37. $9a^3 + 3a^2 - 20a = a(9a^2 + 3a - 20)$

Diagonal product $= -180a^2$, sum $= 3a$
Middle terms are $15a$ and $-12a$.

Factor	$3a$	$+5$
$3a$	$9a^2$	$+15a$
-4	$-12a$	-20

$9a^3 + 3a^2 - 20a = a(3a - 4)(3a + 5)$

39. $6x^3 - 2x^2 - 8x = 2x(3x^2 - x - 4)$

Diagonal product $= -12x^2$, sum $= -x$
Middle terms are $-4x$ and $3x$.

Factor	$3x$	-4
x	$3x^2$	$-4x$
$+1$	$+3x$	-4

$6x^3 - 2x^2 - 8x = 2x(x + 1)(3x - 4)$

41. $12a^3 - 3a = 3a(4a^2 - 1)$
$$= 3a(2a - 1)(2a + 1)$$

Exercises 6.5

1.

Input: x	Output:3^x
2	$3^2 = 9$
1	$3^1 = 3$
0	$3^0 = 1$
-1	$3^{-1} = \frac{1}{3}$
-2	$3^{-2} = \frac{1}{3^2} = \frac{1}{9}$
-3	$3^{-3} = \frac{1}{3^3} = \frac{1}{27}$

3.

Input: x	Output:4^x
2	$4^2 = 16$
1	$4^1 = 4$
0	$4^0 = 1$
-1	$4^{-1} = \frac{1}{4}$
-2	$4^{-2} = \frac{1}{4^2} = \frac{1}{16}$
-3	$4^{-3} = \frac{1}{4^3} = \frac{1}{64}$

5. a. $2^{-2} = \frac{1}{2^2} = \frac{1}{4}$

b. $2^{-1} = \frac{1}{2}$

c. $2^0 = 1$

7. a. $(\frac{1}{4})^0 = 1$

b. $(\frac{1}{4})^{-2} = 4^2 = 16$

c. $(\frac{1}{4})^{-1} = 4^1 = 4$

9. a. $(0.5)^{-1} = (\frac{1}{2})^{-1} = 2^1 = 2$

b. $(0.5)^0 = 1$

c. $(0.5)^{-2} = (\frac{1}{2})^{-2} = 2^2 = 4$

11. An exponent of -1 gives the reciprocal of a number.

13. An exponent of zero gives 1 when the base is not equal to zero.

15. Answers will vary. One possibility is $\frac{1}{4}$ since

$$\left(\frac{1}{4}\right)^{-1} = 4 > \frac{1}{4}.$$

17. a. $5^3 \cdot 5^{-7} = 5^{3+(-7)} = 5^{-4}$

b. $6^5 \cdot 6^{-2} = 6^{5+(-2)} = 6^3$

c. $10^3 \cdot 10^{-6} = 10^{3+(-6)} = 10^{-3}$

d. $10^2 \cdot 10^{-7} = 10^{2+(-7)} = 10^{-5}$

19. a. $10^{-15} \cdot 10^{-15} = 10^{-15+(-15)} = 10^{-30}$

b. $10^{-28} \cdot 10^{19} = 10^{-28+19} = 10^{-9}$

c. $2^{24} \cdot 2^{-16} = 2^{24+(-16)} = 2^8$

d. $2^{13} \cdot 2^{-5} = 2^{13+(-5)} = 2^8$

21. a. $\frac{3^5}{3^{-2}} = 3^{5-(-2)} = 3^7$

b. $\frac{10^{-5}}{10^{-2}} = 10^{-5-(-2)} = 10^{-3}$

c. $\frac{10^{-4}}{10^{12}} = 10^{-4-12} = 10^{-16}$

d. $\frac{1}{10^2} = 10^{-2}$

23. a. $(2^3)^4 = 2^{3\cdot4} = 2^{12}$

b. $(2^3)^{-4} = 2^{3(-4)} = 2^{-12}$

c. $(10^5)^2 = 10^{5\cdot2} = 10^{10}$

d. $(10^{-6})^3 = 10^{(-6)\cdot3} = 10^{-18}$

25. a. $m^3 m^5 = m^{3+5} = m^8$

b. $n^4 n^4 = n^{4+4} = n^8$

c. $a^6 a^2 = a^{6+2} = a^8$

d. $a^7 a^1 = a^{7+1} = a^8$

e. $a^5 \cdot a^{-12} = a^{5+(-12)} = a^{-7}$

f. $x^{-5} \cdot x^{13} = x^{-5+13} = x^8$

g. $n^{-6} \cdot n^2 = x^{-6+2} = n^{-4}$

27. a. $\frac{1}{x^2} = x^{-2}$

b. $\frac{1}{a^{-1}} = a$

c. $\frac{b}{b^{-2}} = b^{1-(-2)} = b^3$

d. $\frac{a^3}{a^{-6}} = a^{3-(-6)} = a^9$

e. $\frac{a^{-6}}{a^2} = a^{-6-2} = a^{-8}$

f. $\frac{x^4}{x^{-2}} = x^{4-(-2)} = x^6$

29. a. $(x^2)^3 = x^{2(3)} = x^6$

b. $(xy)^2 = x^2 y^2$

c. $(x^2 y^3)^2 = (x^2)^2 \cdot (y^3)^2$
$$= x^{2(2)} y^{3(2)}$$
$$= x^4 y^6$$

d. $(x^2)^{-4} = x^{2\cdot(-4)} = x^{-8}$

e. $(x^{-4})^{-3} = x^{-4\cdot(-3)} = x^{12}$

f. $(b^{-2})^3 = b^{-2\cdot3} = b^{-6}$

31. a. $\frac{x^5}{x^2} = x^{5-3} = x^2$

b. $\frac{a^8}{a^5} = a^{8-5} = a^3$

c. $\left(\dfrac{x}{y}\right)^2 = \dfrac{x^2}{y^2}$

d. $\left(\dfrac{2x}{y}\right)^3 = \dfrac{2^3 \cdot x^3}{y^3} = \dfrac{8x^3}{y^3}$

e. $\dfrac{-6b}{9b} = \dfrac{-2 \cdot 3b}{3 \cdot 3b} = -\dfrac{2}{3}$

f. $\dfrac{x^5 y^2}{xy^2} = \dfrac{x^4 \cdot xy^2}{xy^2} = x^4$

g. $\dfrac{-2a^7 b^3}{-8ab^2} = \dfrac{-2ab^2 \cdot a^6 b}{-2ab^2 \cdot 4} = \dfrac{a^6 b}{4}$

33. a. $x^{-1} = \dfrac{1}{x}$

b. 0.0000016

c. $\left(\dfrac{y}{x}\right)^{-1} = \dfrac{x}{y}$

d. $\left(\dfrac{a}{b}\right)^0 = 1$

e. $\left(\dfrac{a}{c}\right)^0 = 1$

f. $\left(\dfrac{a}{bc}\right)^{-1} = \dfrac{bc}{a}$

35. a $y^{-3} = \dfrac{1}{y^3}$

b. $\left(\dfrac{y}{x}\right)^{-2} = \left(\dfrac{x}{y}\right)^2 = \dfrac{x^2}{y^2}$

c. $\dfrac{1}{b^{-3}} = b^3$

d. $\left(\dfrac{a}{b}\right)^{-3} = \left(\dfrac{b}{a}\right)^3 = \dfrac{b^3}{a^3}$

e. $\left(\dfrac{4a^2}{c}\right)^{-2} = \left(\dfrac{c}{4a^2}\right)^2$

$= \dfrac{c^2}{4^2 (a^2)^2}$

$= \dfrac{c^2}{16a^4}$

f. $\left(\dfrac{a}{b^2}\right)^{-3} = \left(\dfrac{b^2}{a}\right)^3 = \dfrac{b^{2 \cdot 3}}{a^3} = \dfrac{b^6}{a^3}$

37. a. $\dfrac{xy^{-2}}{x^2 y^3} = x^{1-2} y^{-2-3} = x^{-1} y^{-5} = \dfrac{1}{xy^5}$

b. $\dfrac{x^{-1} y}{x^{-1} y^{-2}} = x^{-1-(-1)} y^{1-(-2)}$

$= x^0 y^3$

$= y^3$

c. $\dfrac{a^{-2} b^2}{a^3 b^{-1}} = a^{-2-3} b^{2-(-1)}$

$= a^{-5} b^3$

$= \dfrac{b^3}{a^5}$

d. $\dfrac{x^{-2} y}{x^3 y^2} = x^{-2-3} y^{1-2}$

$= x^{-5} y^{-1}$

$= \dfrac{1}{x^5 y}$

e. $\dfrac{x^2 y^{-1}}{x^{-2} y^{-3}} = x^{2-(-2)} y^{-1-(-3)} = x^4 y^2$

f. $\dfrac{a^3 b^{-1}}{a^{-1} b^2} = a^{3-(-1)} b^{-1-2} = a^4 b^{-3} = \dfrac{a^4}{b^3}$

39. a. 1 right; 346

b. 6 left; 0.0000016

c. 5 right; 160,000

d. 1 left; 21.91

e. 3 left; 0.2191

f. 4 right; 2,191,000

41.

x	10^x as a fraction	10^x as a decimal
0	1	1
−1	$\dfrac{1}{10}$	0.1
−2	$\dfrac{1}{100}$	0.01
−3	$\dfrac{1}{1000}$	0.001
−4	$\dfrac{1}{10,000}$	0.0001
−5	$\dfrac{1}{100,000}$	0.00001

43. a. $5280^3 \approx 1.47 \times 10^{11}$

b. $0.00008^3 = 5.12 \times 10^{-13}$

45. $1,391,400 \approx 1.39 \times 10^6$ km

47. 2756.4 million $= 2756.4 \times 10^6$
$\approx 2.76 \times 10^9$ miles

49. $1800 = 1.80 \times 10^3$ g

51. 1,990,000,000,000,000,000,000,000,000,000 kg

53. -0.00000000000000000001602 Coulombs

55. 200,000,000 years

57. 0.0000000000000000000000000016750 kg

59. a. $(2 \times 10^{15})(3 \times 10^{12})$
$= (2)(3) \times 10^{15+12}$
$= 6 \times 10^{27}$

b. $(4 \times 10^{13})(3 \times 10^{18})$
$= (4)(3) \times 10^{13+18}$
$= 12 \times 10^{31} = 1.2 \times 10^{32}$

61. a. $(5 \times 10^{-16})(6 \times 10^{-11})$
$= (5)(6) \times 10^{-16+(-11)}$
$= 30 \times 10^{-27}$
$= 3 \times 10^{-26}$

b. $(8 \times 10^{-12})(6 \times 10^{-13})$
$= (8)(6) \times 10^{-12+(-13)} = 48 \times 10^{-25}$
$= 4.8 \times 10^{-24}$

63. a. $(2.8 \times 10^{-14}) \div (7 \times 10^{-18})$
$= (2.8 \div 7) \times 10^{-14-(-18)} = 0.4 \times 10^4$
$= 4 \times 10^3$

b. $(9 \times 10^{-25}) \div (4.5 \times 10^{-12})$
$= (9 \div 4.5) \times 10^{-25 -(-12)} = 2 \times 10^{-13}$

65. $\dfrac{4.8 \times 10^{-7}}{(1.6 \times 10^8)(3.0 \times 10^{-18})}$
$= \dfrac{4.8 \times 10^{-7}}{4.8 \times 10^{-10}} = 1 \times 10^{-7-(-10)} = 1 \times 10^3$

67. a. $2.34\text{E} - 02 = 2.34 \times 10^{-02} = 0.0234$

b. $3.14\text{E}3 = 3.14 \times 10^3 = 3140$

c. $6.28\text{E}7 = 6.28 \times 10^7 = 62,800,000$

69. a. $34 \times 10^3 = 3.4 \times 10 \times 10^3$
$= 3.4 \times 10^{1+3}$
$= 3.4 \times 10^4$

b. $560 \times 10^{-2} = 5.60 \times 10^2 \times 10^{-2}$
$= 5.60 \times 10^{2+(-2)}$
$= 5.60 \times 10^0$

71. a. $0.432 \times 10^4 = 4.32 \times 10^{-1} \times 10^4$
$= 4.32 \times 10^{-1+4}$
$= 4.32 \times 10^3$

b. $0.567 \times 10^{-5} = 5.67 \times 10^{-1} \times 10^{-5}$
$= 5.67 \times 10^{-1+(-5)}$
$= 5.67 \times 10^{-6}$

73. a. Both numbers are in scientific notation and $10^5 > 10^4$ so,
$\left(2 \times 10^5\right) > \left(3 \times 10^4\right)$

b. Both numbers are in scientific notation and $10^3 < 10^4$
$\left(4 \times 10^3\right) < \left(3 \times 10^4\right)$

75. a. Both numbers are in scientific notation and $10^{-5} < 10^{-4}$ so,
$\left(2 \times 10^{-5}\right) < \left(3 \times 10^{-4}\right)$

b. Both numbers are in scientific notation and $10^{-3} > 10^{-4}$ so,
$\left(4 \times 10^{-3}\right) > \left(3 \times 10^{-4}\right)$

77. a. $3.2 > 2.8$ and $10^{-14} = 10^{-14}$ so,
$$\left(3.2 \times 10^{-14}\right) > \left(2.8 \times 10^{-14}\right)$$

b. $4.3 < 5.3$ and $10^{-13} = 10^{-13}$ so,
$$\left(4.3 \times 10^{-13}\right) < \left(5.3 \times 10^{-13}\right)$$

79. a. $3.2 > -3.2$ and $10^5 = 10^5$ so,
$$\left(3.2 \times 10^5\right) > \left(-3.2 \times 10^5\right)$$

b. $-1.2 = -1.2$ and $10^{-2} < 10^2$.
Since both numbers are negative, the number with the largest absolute value is actually smaller since it lies further to the left along a number line. Thus,
$$\left(-1.2 \times 10^{-2}\right) > \left(-1.2 \times 10^2\right)$$

81. The electron has a smaller mass.
$9.101 \times 10^{-31} < 1.6726 \times 10^{-27}$

83. $\dfrac{186,000 \text{ miles}}{1 \text{ sec}}$

$= \dfrac{1.86 \times 10^5 \text{ miles}}{1 \text{ sec}} \cdot \dfrac{60 \text{ sec}}{1 \text{ min}}$

$= \dfrac{1.116 \times 10^7 \text{ miles}}{1 \text{ min}} \cdot \dfrac{60 \text{ min}}{1 \text{ hr}}$

$= \dfrac{6.696 \times 10^8 \text{ miles}}{1 \text{ hr}} \cdot \dfrac{24 \text{ hr}}{1 \text{ day}}$

$= \dfrac{1.60704 \times 10^{10} \text{ miles}}{1 \text{ day}} \cdot \dfrac{365 \text{ days}}{1 \text{ yr}}$

$\approx 5.87 \times 10^{12}$ miles per year

85. From problem 83 we know that light travels 5.87×10^{12} miles in one light year.
$$\left(5.87 \times 10^{12}\right)\left(2.7 \times 10^4\right)$$
$$= (5.87)(2.7) \times 10^{12+4}$$
$$= 15.849 \times 10^{16}$$
$$\approx 1.58 \times 10^{17} \text{ miles}$$

87. $18,700 \text{ tons} \cdot \dfrac{2000 \text{ lb}}{1 \text{ ton}} \cdot \dfrac{1 \text{ gal}}{8.3 \text{ lb}} \cdot \dfrac{4 \text{ qt}}{1 \text{ gal}}$

$\approx 1.80 \times 10^7$ qts of water

89. 5 ft 9 inches = 69 inches
69 - 20 = 49 inches growth
$$\dfrac{49 \text{ in}}{18 \text{ yr}} \cdot \dfrac{1 \text{ yr}}{365 \text{ days}} \cdot \dfrac{1 \text{ day}}{24 \text{ hr}} \cdot \dfrac{1 \text{ ft}}{12 \text{ in}} \cdot \dfrac{1 \text{ mile}}{5280 \text{ ft}}$$
$\approx 4.90 \times 10^{-9}$ mph

91. The bases are the same in $a^5 \cdot a^8$ but not in $a^2 b^2$.

93. Multiply the two powers.
$$\left(a^4\right)^3 = a^{4(3)} = a^{12}$$

95. Written response. See page 358 in text.

Chapter 6 Review Exercises

1. a. $3a^2 - 5ab + 4a^2 - 3b^2 + 2ab - 7b^2$
$= (3 + 4)a^2 + (-5 + 2)ab + (-3 - 7)b^2$
$= 7a^2 - 3ab - 10b^2$, trinomial

b. $3 + 4x^2 + 5x - 2(x - 1)$
$= 4x^2 + 5x - 2x + 2 + 3$
$= 4x^2 + 3x + 5$, trinomial

c. $x(x^2 + 4x + 16) - 4(x^2 + 4x + 16)$
$= x^3 + 4x^2 + 16x - 4x^2 - 16x - 64$
$= x^3 - 64$, binomial

d. $11x - 4x(1 - x) = 11x - 4x + 4x^2$
$= 4x^2 + 7x$, binomial

e. $9 - 4(x - 3) = 9 - 4x + 12$
$= -4x + 21$, binomial

3. a. $P = 2(2x) + 2(3x + 1)$,
$P = 4x + 6x + 2$,
$P = 10x + 2$
$A = 2x(3x + 1)$, $A = 6x^2 + 2x$

b. $P = 2(x + 2) + 2(3x)$,
$P = 2x + 4 + 6x$,
$P = 8x + 4$
$A = 3x(x + 2)$, $A = 3x^2 + 6x$

c. $P = (3x - 4) + 2x + (2x + 1)$
$P = 3x + 2x + 2x - 4 + 1$,
$P = 7x - 3$
$A = \frac{1}{2}(2x+1)(2x)$, $A = (2x+1)x$,
$A = 2x^2 + x$

5. $(3x + 1)(x + 1) = 3x^2 + 4x + 1$

7. $(x + 4)(x + 3) = x^2 + 3x + 4x + 12$
$= x^2 + 7x + 12$

9. $(2x - 5)(2x - 5)$
$= (2x)^2 + 2(2x)(-5) + (-5)^2$
$= 4x^2 - 20x + 25$

11. $(3x + 5)(3x - 2)$
$= 9x^2 - 6x + 15x - 10$
$= 9x^2 + 9x - 10$

13. $(2x - 3)(3x + 2) = 6x^2 + 4x - 9x - 6$
$= 6x^2 - 5x - 6$

15. $(a - b)(a - b) = a^2 - 2ab + b^2$

17. Exercises 9 and 15 are perfect square trinomials.

19. a.

Factor	x	+7
x	x^2	+7x
-2	-2x	-14

$(x + 7)(x - 2) = x^2 + 5x - 14$

b.

Factor	3x	+5
2x	$6x^2$	+10x
+3	+9x	+15

$(3x + 5)(2x + 3) = 6x^2 + 19x + 15$

21. a. Diagonal product $= 14x^2$,
sum $= -9x$

Factor	x	-2
x	x^2	-2x
-7	-7x	+14

$(x - 2)(x - 7) = x^2 - 9x + 14$

b. Diagonal product $= -42x^2$,
sum $= 11x$,

Factor	x	+7
2x	$2x^2$	+14x
-3	-3x	-21

$(x + 7)(2x - 3) = 2x^2 + 11x - 21$

c. Diagonal product $= -12x^2$,
sum $= -x$,

Factor	3x	-1
4x	$12x^2$	-4x
+1	+3x	-1

$(3x - 1)(4x + 1) = 12x^2 - x - 1$

d. Diagonal product $= -180x^2$,
sum $= 24x$

Factor	2x	+3
10x	$20x^2$	+30x
-3	-6x	-9

$(2x + 3)(10x - 3) = 20x^2 + 24x - 9$

23. $x^2 - 3x + 2 = x^2 - x - 2x + 2$
$= (x - 1)(x - 2)$, neither

25. $9x^2 - 16 = (3x + 4)(3x - 4)$, ds

27. $2x^2 + x - 6 = 2x^2 + 4x - 3x - 6$
$= (2x - 3)(x + 2)$, neither

29. $9x^2 + 3x - 2 = 9x^2 + 6x - 3x - 2$
$= (3x + 2)(3x - 1)$, neither

31. $x^2 + 6x + 8 = x^2 + 4x + 2x + 8$
$= (x + 4)(x + 2)$, neither

33. $x^2 - 11x + 10 = x^2 - x - 10x + 10$
$= (x - 1)(x - 10)$, neither

35. $25 - 9x^2 = (5 + 3x)(5 - 3x)$, ds

37. $x^2 - 4x + 6$ does not factor.

39. $y^2 + 8y + 12 = y^2 + 6y + 2y + 12$
$= (y + 6)(y + 2)$, neither

41. $2x^2 - 3x - 35 = 2x^2 - 10x + 7x - 35$
$= (2x + 7)(x - 5)$, neither

43. $4x^2 - 8x + 4 = 4(x^2 - 2x + 1)$
$= 4(x - 1)^2$, pst

45. $x^3 + 4x^2 + 4x = x(x^2 + 4x + 4)$
$= x(x + 2)^2$, contains pst

47. $3x^2 - 27 = 3(x^2 - 9)$
$= 3(x + 3)(x - 3)$, contains ds

49. $x^3 - 7x^2 + 10x = x(x^2 - 7x + 10)$
$= x(x^2 - 2x - 5x + 10) = x(x - 2)(x - 5)$, neither

51. $2x^2 + 8x + 14 = 2(x^2 + 4x + 7)$;
neither

53. a. $3^{-1} = \frac{1}{3}$

b. $3^0 = 1$

c. $3^{-2} = \frac{1}{3^2} = \frac{1}{9}$

d. $\left(\frac{2}{3}\right)^0 = 1$

e. $\left(\frac{2}{3}\right)^{-1} = \frac{3}{2}$

f. $\left(\frac{2}{3}\right)^{-2} = \left(\frac{3}{2}\right)^2 = \frac{9}{4}$

55. a. $x^7 x^{-2} = x^{7+(-2)} = x^5$

b. $x^3 x^{-3} = x^{3+(-3)} = x^0 = 1$

c. $b^{-5} b^{-5} = b^{-5+(-5)} = \frac{1}{b^{10}}$

57. a. $\frac{n^4}{n^{-5}} = n^{4-(-5)} = n^9$

b. $\frac{n^{-4}}{n^{-5}} = n^{-4-(-5)} = n$

c. $n^{-9} = \frac{1}{n^9}$

59. a. $(b^2)^{-4} = b^{2\cdot(-4)} = b^{-8} = \frac{1}{b^8}$

b. $(x^{-2})^{-3} = x^{(-2)(-3)} = x^6$

c. $(b^0)^{-2} = b^{0\cdot(-2)} = b^0 = 1$

61. a. $\left(\frac{a^2}{b^2}\right)^{-1} = \frac{b^2}{a^2}$

b. $\left(\frac{2a}{b^2c}\right)^{-3} = \left(\frac{b^2c}{2a}\right)^3 = \frac{b^6c^3}{8a^3}$

c. $\left(\frac{3b}{a^2}\right)^0 = 1$

63. a. $\frac{x^{-3}y^{-3}}{x^5 y^{-2}} = x^{-3-5} y^{-3-(-2)}$
$= x^{-8} y^{-1}$
$= \frac{1}{x^8 y}$

b. $\frac{a^3 b^{-6}}{a^6 b^{-4}} = a^{3-6} b^{-6-(-4)} = a^{-3} b^{-2}$
$= \frac{1}{a^3 b^2}$

65. a. $\frac{x^3 x^0}{x^0} = x^{3+0-0} = x^3$

b. $\frac{a^{-2} a^2}{a^0} = a^{-2+2-0} = a^0 = 1$

67.

Year	National Debt	Population	Debt per person
1900	1.2×10^9	7.62×10^7	15.7
1920	2.42×10^{10}	1.06×10^8	228
1950	2.561×10^{11}	1.513×10^8	1690
1990	3.2333×10^{12}	2.487×10^8	13,000
2000	5.6742×10^{12}	2.814×10^8	20,200

69. 1.243×10^{-7} years $\cdot \frac{365\ \text{days}}{1\ \text{year}}$

$\approx 4.537\times10^{-5}$ days $\cdot \frac{24\ \text{hr}}{1\ \text{day}}$

$\approx 1.089\times10^{-3}$ hrs $\cdot \frac{60\ \text{min}}{1\ \text{hr}}$

≈ 0.0653 min $\cdot \frac{60\ \text{sec}}{1\ \text{min}}$

≈ 3.92 sec

71. a. $(1.5\times10^{-4})(3.0\times10^9)$
$= (1.5)(3.0)\times10^{-4+9}$
$= 4.5\times10^5$

b. $\left(2.5\times10^{-9}\right)\left(4.0\times10^{3}\right)$

$=(2.5)(4.0)\times10^{-9+3}$

$=10\times10^{-6}$

$=1.0\times10\times10^{-6}$

$=1.0\times10^{-5}$

73. a. $\dfrac{6.4\times10^{-12}}{1.6\times10^{-3}}=\dfrac{6.4}{1.6}\times\dfrac{10^{-12}}{10^{-3}}$

$=4.0\times10^{-12-(-3)}$

$=4.0\times10^{-9}$

b. $\dfrac{7.5\times10^{-13}}{2.5\times10^{-6}}=\dfrac{7.5}{2.5}\times\dfrac{10^{-13}}{10^{-6}}$

$=3.0\times10^{-13-(-6)}$

$=3.0\times10^{-7}$

75. a. $38.5\times10^{-3}=3.85\times10\times10^{-3}$

$=3.85\times10^{1+(-3)}$

$=3.85\times10^{-2}$

b. $0.48\times10^{-2}=4.8\times10^{-1}\times10^{-2}$

$=4.8\times10^{-1+(-2)}$

$=4.8\times10^{-3}$

77. $\dfrac{1.5\times10^{6}\text{ miles}}{27.3\text{ days}}\cdot\dfrac{1\text{ day}}{24\text{ hours}}$

≈2289.38 miles per hour

The speed of the moon is roughly 2290 miles per hour.

Chapter 6 Test

1. 3 terms, $3x^2$, $4x$ and 1

2. Greatest common factor

3. Factoring

4. -1

5. $4^0=1$, $x=0$

6. $\dfrac{1}{16}=\dfrac{1}{4^2}=4^{-2}$, $x=-2$

7. 0.0003482

8. 4.5×10^{10}

9. $(a+3b-3c-d)+(3a-5b+8c-d)$
$=a+3a+3b-5b-3c+8c-d-d$
$=4a-2b+5c-2d$, 4 term polynomial

10 $x(x^2+3x+9)-3(x^2+3x+9)$
$=x^3+3x^2+9x-3x^2-9x-27$
$=x^3-27$, binomial

11. $-4x(x-5)=-4x^2+20x$, binomial

12. $14xy+6x^2y-18y^2$
$=6x^2y+14xy-18y^2$
$=2y(3x^2+7x-9y)$

13.

Factor	3x	+4
2x	$6x^2$	+8x
-5	-15x	-20

$(3x+4)(2x-5)=6x^2-7x-20$

14. $P=2(4x)+2(2x+5)$,
$P=8x+4x+10, P=12x+10$
$A=4x(2x+5), A=8x^2+20x$

15. Let w = width;
$6x^2+9x=w(2x+3)$,
$3x(2x+3)=w(2x+3)$,
$w=3x$

16. $(x-4)(x+7)=x^2+7x-4x-28$
$=x^2+3x-28$

17. $(x-7)(x-7)=x^2+2(-7x)+(-7)^2$
$=x^2-14x+49$

18. $(2x-7)(2x+7)=(2x)^2-7^2=4x^2-49$

19. $2(x-4)(x-4)=2(x^2-8x+16)$
$=2x^2-16x+32$

20. $x^2-9x+20=x^2-4x-5x+20$
$=(x-4)(x-5)$

21. $2x^2-3x-2=2x^2-4x+x-2$
$=(2x+1)(x-2)$

22. $2x^2-8=2(x^2-4)=2(x+2)(x-2)$

23. $x^2-8x+16=(x-4)^2$

24. $9x^2+6x+1=(3x+1)^2$

25. $3x^2+6x+15=3\left(x^2+2x+5\right)$

26. Exercises 18 and 22 are difference of squares.

27. a. $b^{-2}b^3 = b^{-2+3} = b$

b. $\left(x^3\right)^{-2} = x^{3\cdot(-2)} = x^{-6} = \dfrac{1}{x^6}$

28. a. $\dfrac{b^3}{b^{-2}} = b^{3-(-2)} = b^5$

b. $\dfrac{ab^0}{a^2b^{-1}} = a^{1-2}b^{0-(-1)} = a^{-1}b^1 = \dfrac{b}{a}$

29. a. $\left(\dfrac{a}{2b}\right)^0 = 1$

b. $\left(\dfrac{9x^2}{25y^2}\right)^{-2} = \left(\dfrac{25y^2}{9x^2}\right)^2 = \dfrac{625y^4}{81x^4}$

30. $(2.5 \times 10^{-15})(4.0 \times 10^2)$
$= (2.5)(4.0) \times 10^{-15+2} = 10 \times 10^{-13}$
$= 1 \times 10^{-12} = 0.000000000001$

31. $\dfrac{1.25 \times 10^{-8}}{2.5 \times 10^{-4}} = \dfrac{1.25}{2.5} \times 10^{-8-(-4)} = 0.5 \times 10^{-4}$
$= 5 \times 10^{-5}$

32. Factors
$(x \pm 1)(x \pm 21)$; $(x \pm 3)(x \pm 7)$

Trinomials
$x^2 \pm 22x + 21$, $x^2 \pm 10x + 21$,
$x^2 \pm 20x - 21$, $x^2 \pm 4x - 21$.

1, 21 and 3, 7 are the only factors of 21.

33.

Multiply	a	+b
a	a^2	+ab
+b	+ab	b^2

There are two places in the table where a and b multiply each other resulting in the missing center term of 2ab.

34. 10 EE 3 is really $10 \times 10^3 = 10^{1+3}$
$= 10^4 = 10,000$;
10 ^ 3 is $10^3 = 1000$.

Cumulative Review of Chapters 1 to 6

1.

Input: x	Output: y = 0.25x + 0.25
0	0.25(0) + 0.25 = 0.25
5	0.25(5) + 0.25 = 1.50
10	0.25(10) + 0.25 = 2.75
15	0.25(15) + 0.25 = 4.00
20	0.25(20) + 0.25 = 5.25
25	0.25(25) + 0.25 = 6.50
30	0.25(30) + 0.25 = 7.75

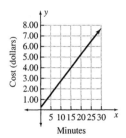

3. $5 - 2(4 - x) = 5 - 8 + 2x = 2x - 3$

5. $x^3x^8 = x^{3+8} = x^{11}$

7. $R = \dfrac{(30)(20)}{30+20}$, $R = \dfrac{600}{50}$, $R = 12$

9. $3x - 13 = 13 - 2x$, $5x - 13 = 13$,
$5x = 26$, $x = 26 \div 5$, $x = 5.2$

11. $x + 1 \geq 9 - 3x$, $4x + 1 \geq 9$,
$4x \geq 8$, $x \geq 2$

$\begin{array}{c} \hline -5\ -4\ -3\ -2\ -1\ \ 0\ \ 1\ \ 2\ \ 3\ \ 4\ \ 5 \end{array}$

13. $3x + 5y = 15$, $5y = -3x + 15$,
$y = \dfrac{-3x + 15}{5}$, $y = -\tfrac{3}{5}x + 3$

15. $\dfrac{a+b+c}{3} = m$, $a + b + c = 3m$
$a = 3m - b - c$

17.

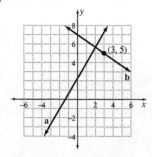

19. Slope from exercise $13 = -\frac{3}{5}$, $b = 0$,

$y = -\frac{3}{5}x$

21. $\dfrac{1.19 - 0.92}{0.92} = \dfrac{0.27}{0.92} \approx 0.293$

29.3% increase

23. $\dfrac{x-1}{2} = \dfrac{x+3}{3}$, $\quad 3(x-1) = 2(x+3)$

$3x - 3 = 2x + 6$, $\quad x = 9$

25. $x(x^2 - 2x + 1) - (x^2 - 2x + 1)$
$= x^3 - 2x^2 + x - x^2 + 2x - 1$
$= x^3 - 3x^2 + 3x - 1$

27. $(2x + 1)(2x + 3) = 4x^2 + 6x + 2x + 3$
$= 4x^2 + 8x + 3$

29. $4x^2 - 25 = (2x + 5)(2x - 5)$, difference of squares

31. 1.36×10^9 lb $\cdot \dfrac{16 \text{ oz}}{1 \text{ lb}} \cdot \dfrac{160 \text{ cal}}{1 \text{ oz}}$

$\approx 3.4816 \times 10^{12}$ cal.
1 billion $= 10^9$, $(3.4816 \times 10^{12}) \div 5.6 \times 10^9$
$= (3.4816 \div 5.6) \times 10^{12-9} \approx 0.6217 \times 10^3$
≈ 622 calories per person

Chapter 7

Section 7.1

1. In $y = -2x$, $m = -2$, $b = 0$
In $y = 1 - x$, $m = -1$, $b = 1$
Intersection is at $(-1, 2)$
Check:
$2 = -2(-1)$, $2 = 2$ ✔
$2 = -1 - (-1)$, $2 = 2$ ✔

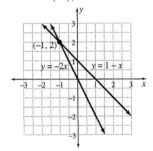

3. In $y = x$, $m = 1$, $b = 0$
In $y = 2x + 3$, $m = 2$, $b = 3$
Intersection is at $(-3, -3)$
Check:
$-3 = -3$, ✔
$-3 = 2(-3) + 3$, $-3 = -3$ ✔

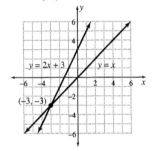

5. In $y = 3$, $m = 0$, $b = 3$
In $x = -1$, m is undefined, no y-intercept
Intersection is at $(-1, 3)$
Check:
$3 = 3$ ✔
$-1 = -1$ ✔

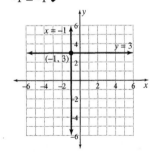

7. Solving $x + y = 5$ for y; $y = -x + 5$, $m = -1$, $b = 5$. Both equations describe the same graph, coincident lines with an infinite number of solutions.

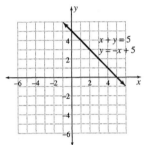

9. Solving $x - y = 3$ for y, $y = x - 3$, $m = 1$, $b = -3$
In $y = x - 6$, $m = 1$, $b = -6$

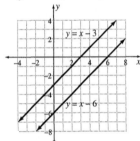

No point of intersection, same slope different y-intercepts, parallel lines.

11. In $y = x - 4$, $m = 1$, $b = -4$
Solving for y, $y = -3x + 4$, $m = -3$, $b = 4$
Intersection is at $(2, -2)$
Check:
$-2 = 2 - 4$, $-2 = -2$ ✔
$-2 = -3(2) + 4$, $-2 = -2$ ✔

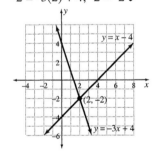

13. Solving for y, $y = 2x - 1$, $m = 2$, $b = -1$
In $y = 4x + 3$, $m = 4$, $b = 3$
Intersection is at $(-2, -5)$
Check:
$-5 = 2(-2) - 1$, $-5 = -5$ ✔
$-5 = 4(-2) + 3$, $-5 = -5$ ✔

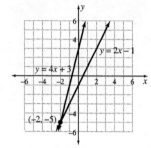

15. Solving for y, $y = -\frac{2}{3}x + 4$,
$m = -\frac{2}{3}$, $b = 4$
Solving for y, $y = -x + 5$, $m = -1$, $b = 5$
Intersection is at $(3, 2)$
Check:
$2 = -\frac{2}{3}(3) + 4$, $2 = 2$ ✔
$2 = -(3) + 5$, $2 = 2$ ✔

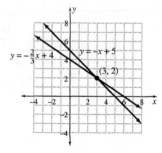

17. $2x + 2y = 100$, $2y = -2x + 100$,
$y = -x + 50$, $m = -1$, $b = 50$
In $y = 20 - x$, $m = -1$, $b = 20$
Same slope, different y-intercepts,
Parallel lines

19. $y = 55x$, $m = 55$, $b = 0$
$y = 25x$, $m = 25$, $b = 0$
Different slopes, not parallel

21. $y = 4x$, $m = 4$, $b = 0$
$y = \frac{1}{4}(12 + 16x)$, $y = 3 + 4x$, $m = 4$, $b = 3$
Same slope, different y-intercepts,
Parallel lines

23. $y + 60x = 300$, $y = -60x + 300$, $m = -60$, $b = 300$
$y = 60 - 300x$, $m = -300$, $b = 60$
Different slopes, different y-intercepts
Not coincident

25. $y = x + 0.15x$, $y = 1.15x$, $m = 1.15$, $b = 0$
$y = 1.15x$, same equation as above,
Same slope, same y-intercept,
Coincident lines.

27. $2x + y = 10$, $y = -2x + 10$, $m = -2$, $b = 10$
$2y + x = 10$, $2y = -x + 10$, $y = -\frac{1}{2}x + 5$,
$m = -\frac{1}{2}$, $b = 5$
Different slopes, different y-intercepts
Not coincident

29. a. One line has y-intercept $= 5$, slope is -2,
equation is $y = -2x + 5$

   b. The other line has y-intercept $= 3$,
slope is -1, equation is $y = -x + 3$.
Thus the system is
$y = -2x + 5$
$y = -x + 3$
Intersection is at $(2, 1)$
Check: $1 = -2(2) + 5$, $1 = 1$ ✔
             $1 = -(2) + 3$, $1 = 1$ ✔

31. One of the lines has slope $m = \frac{1}{3}$ and y-intercept
$b = 2$. The equation of this line is $y = \frac{1}{3}x + 2$.
The other line has slope $m = \frac{5}{3}$ and y-intercept
$b = 6$. The equation of this line is $y = \frac{5}{3}x + 6$.
Thus, the system is
$y = \frac{1}{3}x + 2$
$y = \frac{5}{3}x + 6$
The two lines intersect at the point $(-3, 1)$.

Check: $(1) \overset{?}{=} \frac{1}{3}(-3) + 2$ ✔
           $(1) \overset{?}{=} \frac{5}{3}(-3) + 6$ ✔

33. One of the lines has slope $m = \frac{3}{2}$ and
y-intercept $b = 2$. The equation of this line is
$y = \frac{3}{2}x + 2$. The other line has slope $m = -\frac{3}{2}$
and y-intercept $b = -4$. The equation of this line
is $y = -\frac{3}{2}x - 4$. Thus, the system is

$y = \frac{3}{2}x + 2$

$y = -\frac{3}{2}x - 4$

The two lines intersect at the point $(-2, -1)$.

Check: $(-1) \overset{?}{=} \frac{3}{2}(-2) + 2$ ✔

$(-1) \overset{?}{=} -\frac{3}{2}(-2) - 4$ ✔

35. a. y = 50 + 0.12x rewards big sales because the slope is larger.

b.

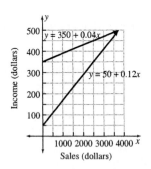

c. Intersection is at (3750, 500)

d. y = 350 + 0.04x would be preferred when sales are less than 3750.

e. Doubling the rate per dollar sales is the same as doubling the slope. New equations are:
y = 350 + 0.08x   and   y = 50 + 0.24x

37. a. From the graph the total cost for 100 people is about $525; using the equation
C = 150 + 3.75(100), C = 525.

b. From the graph the total revenue is $500, or using the equation;
R = 5(100), R = 500.

c. Profit is R - C, 500 - 525 = -25

d. The cost graph is on top when total registration is less than the breakeven point.

39. a. C = $8.50x + $250

b. R = $10x + $200

c.

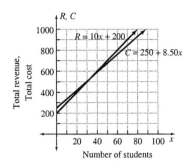

d. From the graph the break-even point is roughly (34, 540).

$8.5(34) + 250 = 539 \approx 540$

$10(34) + 200 = 540$

The actual breakeven point is between 33 and 34; since we must have whole people we take the larger number.

Section 7.2

1.

Long seg.	Short seg.	Total (m)
1 + 5 = 6	1	6 + 1 = 7
2 + 5 = 7	2	7 + 2 = 9
3 + 5 = 8	3	8 + 3 = 11
x	y	10

The total length is 10 meters, so
x + y = 10.
One piece is 5 meters longer than the other, so
x = y + 5.
Thus, the system of equations is
$x + y = 10$
$x = y + 5$

3.

Project A	Project B	Total
hr	hr	hr
10	10 + 28 = 38	10 + 38 = 48
20	20 + 28 = 48	20 + 48 = 68
50	50 + 28 = 78	50 + 78 = 128
x	y	176

The total number of hours worked is 176, so
$x + y = 176$.
He works 28 hours longer on one project than
another, so $y = x + 28$.
Thus, the system of equations is
$x + y = 176$

$y = x + 28$

5. As a guess, choose 10 nickels and 12 quarters.

Item	Quantity	Value	$Q \cdot V$
Nickels	10	0.05	0.50
Quarters	12	0.25	3.00
Total	22		3.50

The number of coins is 22, but the value of the
coins is $3.50, not $2.90.

Let x represent the number of nickels and let y
represent the number of quarters.

Item	Quantity	Value	$Q \cdot V$
Nickels	x	0.05	$0.05x$
Quarters	y	0.25	$0.25y$
Total	22		2.90

The system of equations is
$x + y = 22$

$0.05x + 0.25y = 2.90$

7. As a guess, choose 14 dimes and 12 quarters.

Item	Quantity	Value	$Q \cdot V$
Dimes	14	0.10	1.40
Quarters	12	0.25	3.00
Total	26		4.40

The number of coins is 26, but the value of the
coins is $4.40, not $5.40.

Let x represent the number of dimes and let y
represent the number of quarters.

Item	Quantity	Value	$Q \cdot V$
Dimes	x	0.10	$0.10x$
Quarters	y	0.25	$0.25y$
Total	26		5.45

The system of equations is
$x + y = 26$

$0.1x + 0.25y = 5.45$

9. As a guess, choose 40 nickels and 25 dimes.

Item	Quantity	Value	$Q \cdot V$
Nickels	40	0.05	2.00
Dimes	25	0.10	2.50
Total	65		4.50

The number of coins is 65, but the value of the
coins is $4.50, not $5.40.

Let x represent the number of nickels and let y
represent the number of quarters.

Item	Quantity	Value	$Q \cdot V$
Nickels	x	0.05	$0.05x$
Dimes	y	0.10	$0.10y$
Total	65		5.40

The system of equations is
$x + y = 65$

$0.05x + 0.10y = 5.45$

11. As a guess, choose 20 hours at the 1st job and 23
hours at the second job.

Item	Quantity (hours)	Value ($/hr)	$Q \cdot V$ ($)
1st job	20	5.75	115
2nd job	23	8.50	195.50
Total	43		310.50

The number of hours is 43, but the total wages is
$310.50, not $316.

Let x represent the number of hour at the 1st job
and let y represent the number of hours at the
second job.

Item	Quantity (hours)	Value ($/hr)	$Q \cdot V$ ($)
1st job	x	5.75	$5.75x$
2nd job	y	8.50	$8.50y$
Total	43		316

The system of equations is
$x + y = 43$

$5.75x + 8.50y = 316$

13. As a guess, choose 750 pounds of carrots to restaurants and 4500 pounds of carrots to grocery stores.

Restaurant lbs	Grocery lbs	Total lbs	Groc. / Rest.
750	4500	5250	6

The number of pounds is 5250, but the pounds to grocery stores is only 6 times as much as to restaurants, not 9 times.

Let y represent the number of pounds to restaurants and x represent the number of pounds to grocery stores.

Restaurant lbs	Grocery lbs	Total lbs	Groc. / Rest.
y	x	5250	9

The system of equations is
$x + y = 5250$

$\frac{x}{y} = 9$, $x = 9y$

15. Let x represent the price of an adult's ticket and let y represent the price of a student's ticket. The system of equations is
$2x + 3y = 38.50$

$3x + 2y = \$41.50$

Use guess and check to solve. The price of an adult ticket is \$9.50 and the price of a student ticket is \$6.50.

$2(9.50) + 3(6.50) \overset{?}{=} 38.5$ ✔

$3(9.50) + 2(6.50) \overset{?}{=} 41.50$ ✔

17. Let x represent the price of a kids ticket and let y represent the price of an adult ticket. The system of equations is
$5x + 2y = 30$

$6x + y = 32.50$

Use guess and check to solve. The price of a kids ticket is \$5.00 and the price of an adult ticket is \$2.50.

$5(5.00) + 2(2.50) \overset{?}{=} 30$ ✔

$6(5.00) + (2.50) \overset{?}{=} 32.50$ ✔

19. Let x be the top price and let y be the lowest-price. The system of equations is:
$2x + 5y = 800$

$3x + 3y = 1020$

Use guess and check to solve. The top price ticket costs \$300 and the lowest price ticket costs \$40.

$2(300) + 5(40) \overset{?}{=} 800$ ✔

$3(300) + 3(40) \overset{?}{=} 1020$ ✔

21. a. $8 - 3 = 5$ mph

b. $8 - 6 = 2$ mph

c. $8 - 15 = -7$ mph

d. $8 + 3 = 11$ mph

e. $8 + 5 = 13$ mph

23. Let r represent the airspeed of the plane and let w represent the wind speed.

	Net rate	Time	Distance
With wind	$r + w$	5	1400
Against wind	$r - w$	5.6	1400

The system of equations is
$5(r + w) = 1400$

$5.6(r - w) = 1400$

Use guess and check to solve. The airspeed of the plane is 265 mph and the wind speed is 15 mph.

$5(265 + 15) \overset{?}{=} 1400$ ✔

$5.6(265 - 15) \overset{?}{=} 1400$ ✔

25. Let r represent the speed of the boat in still water, and let c represent the speed of the current.

	Net rate	Time	Distance
Upstream	$r - c$	1.2	12
Downstream	$r + c$	0.8	12

The system of equations is

$$1.2(r-c)=12$$

$$0.8(r+c)=12$$

Use guess and check to solve. The speed of the boat is 12.5 mph and the speed of the current is 2.5 mph.

$$1.2(12.5-2.5)\overset{?}{=}12 \ \checkmark$$

$$0.8(12.5+2.5)\overset{?}{=}12 \ \checkmark$$

Section 7.3

1. $L = 2W$,

   $W = \dfrac{L}{2}$

3. $a + b = c$,

   $b = c - a$

5. $C = 2\pi r$,

   $r = \dfrac{C}{2\pi}$

7. $x - y = 5$,

   $x = 5 + y$,

   $y = x - 5$

9. $C = \pi d$,

   $d = \dfrac{C}{\pi}$

11. $2x + 3(2) = 12$,

    $2x + 6 = 12$,

    $2x = 6$,

    $x = 3$

13. $-5x - 6(-3) = 3$,

    $-5x + 18 = 3$

    $18 = 3 + 5x$

    $15 = 5x$

    $x = 3$

15. $y, 3x + y = 4$,

    $y = 4 - 3x$

17. $x, x - 4y = 5$,

    $x = 5 + 4y$

19. $x, 5y - x = 9$,

    $5y = 9 + x$,

    $x = 5y - 9$

21. $y, 3x - y = -2$,

    $3x = -2 + y$,

    $y = 3x + 2$

23. Substitute $y = x - 8$ into $3x + y = 4$,

    $3x + (x - 8) = 4$,

    $4x - 8 = 4$,

    $4x = 12$,

    $x = 3$

    Now substitute $x = 3$ into $y = x - 8$,

    $y = 3 - 8$,

    $y = -5$

    Check your solutions:

    $-5 = 3 - 8$?

    $3(3) + (-5) = 4$?

25. Substitute $y = 5x + 5$ into $y - 3x = 9$,

    $(5x + 5) - 3x = 9$,

    $2x + 5 = 9$,

    $2x = 4$,

    $x = 2$

    Now substitute $x = 2$ into $y = 5x + 5$,

    $y = 5(2) + 5$,

    $y = 15$

    Check your solutions:

    $15 = 5(2) + 5$?

    $15 - 3(2) = 9$?

27. Substitute $x = 9 + 5y$ into $4x + 5y = 11$,

    $4(9 + 5y) + 5y = 11$,

    $36 + 20y + 5y = 11$,

    $36 + 25y = 11$,

    $25y = -25$,

    $y = -1$

    Now substitute $y = -1$ into $x = 9 + 5y$,

    $x = 9 + 5(-1)$,

    $x = 4$

    Check your solutions:

    $4 = 9 + 5(-1)$?

    $4(4) + 5(-1) = 11$?

29. First solve $2x - y = 1$ for y,

    $2x - y = 1$,

    $2x = 1 + y$,

    $y = 2x - 1$

    Substitute for y in $2y = 3x + 3$,

    $2(2x - 1) = 3x + 3$,

    $4x - 2 = 3x + 3$,

    $x - 2 = 3$,

    $x = 5$

Now substitute $x = 5$ into $y = 2x - 1$,
$y = 2(5) - 1$,
$y = 9$
Check your solutions:
$2(5) - 9 = 1$?
$2(9) = 3(5) + 3$?

31. Solve $3x + y = 7$ for y,
$y = 7 - 3x$
Substitute into $2x + 3y = 0$,
$2x + 3(7 - 3x) = 0$,
$2x + 21 - 9x = 0$,
$-7x = -21$,
$x = 3$

Substitute into $y = 7 - 3x$,
$y = 7 - 3(3)$,
$y = -2$
Check your solutions,
$2(3) + 3(-2) = 0$?
$3(3) + (-2) = 7$?

33. Both equations are solved for y, set them equal to each other,
$\frac{4}{3}x = -\frac{8}{3}x + 8$,
$4x = -8x + 24$,
$12x = 24$,
$x = 2$

Substitute into $y = \frac{4}{3}x$,
$y = \frac{4}{3}(2)$
$y = \frac{8}{3}$
Check your solutions,
$\frac{8}{3} = \frac{4}{3}(2)$?
$\frac{8}{3} = -\frac{8}{3}(2) + 8$?

35. Substitute $y = 3x + 4$ into $3x - y = 8$,
$3x - (3x + 4) = 8$,
$3x - 3x - 4 = 8$,
$-4 = 8$, this is a false statement, the system of equations has no solution.

37. Substitute $y = 2x - 3$ into $y - 2x = 5$,
$(2x - 3) - 2x = 5$,
$2x - 3 - 2x = 5$,
$-3 = 5$, this is a false statement, the system of equations has no solution.

39. Substitute $x = 7 - y$ into $x + y = 7$,
$(7 - y) + y = 7$.
$7 = 7$, this is always true, the system of equations has an infinite number of solutions.

41. Set equations equal to each other,
$3x + 2 = -x - 2$,
$4x + 2 = -2$,
$4x = -4$,
$x = -1$

Substitute into one of the equations,
$y = -(-1) - 2$,
$y = 1 - 2$,
$y = -1$
Check your solutions,
$-1 = 3(-1) + 2$?
$-1 = -(-1) - 2$?

43. Set equations equal to each other,
$-x - 2 = 3x - 2$,
$-2 = 4x - 2$,
$0 = 4x$,
$x = 0$

Substitute into one of the equations,
$y = -(0) - 2$.
$y = -2$
Check your solutions,
$-2 = -(0) - 2$?
$-2 = 3(0) - 2$?

45. Solve $2y - x = -10$ for x, $x = 2y + 10$
Substitute into $2x + 3y = 6$,
$2(2y + 10) + 3y = 6$,
$4y + 20 + 3y = 6$,
$7y = -14$,
$y = -2$

Substitute into $x = 2y + 10$,
$x = 2(-2) + 10$,
$x = 6$
Check your solutions,
$2(-2) - (6) = -10$?
$2(6) + 3(-2) = 6$?

47. Substitute $y = \frac{1}{2}x - 2$ into $2y + 4 = x$,
$2(\frac{1}{2}x - 2) + 4 = x$,
$x - 4 + 4 = x$,
$x = x$, this is always true, system of equations has an infinite number of solutions.

49. Solve $x + y = 50$ for y, $y = 50 - x$,
Substitute into $7.50x + 9.75y = 435.75$,
$7.50x + 9.75(50 - x) = 435.75$,
$7.50x + 487.50 - 9.75x = 435.75$,
$487.50 - 2.25x = 437.75$,
$51.75 = 2.25x$,
$x = 23$

Substitute into y = 50 - x,
y = 50 - 23,
y = 27
Check your solutions,
23 + 27 = 50?
7.50(23) + 9.75(27) = 435.75?

51. Set equations equal to each other,
45 + 5(x - 10) = 30 + 3x,
45 + 5x - 50 = 30 + 3x,
5x - 5 = 30 + 3x,
2x = 35,
x = 17.5

Substitute into one of the equations,
y = 30 + 3(17.5),
y = 82.5
Check your solutions,
82.5 = 45 + 5(17.5 - 10)?
82.5 = 30 + 3(17.5)?

53. Let x and y represent the 2 pieces of ribbon, with y being the longer piece.
x + y = 20, y = x + 3
x + (x + 3) = 20,
2x + 3 = 20,
2x = 17,
x = 8.5
y = 8.5 + 3, y = 11.5
The ribbons are 8.5 yd and 11.5 yd long.

55. Let h = height and w = width.
Recall perimeter is 2h + 2w
40 = 2h + 2w, h = 4 + w,
40 = 2(4 + w) + 2w,
40 = 8 + 2w + 2w
32 = 4w,
w = 8
h = 4 + 8, h = 12
Height is 12 in, width is 8 in.

57. Let h = height and w = width.
58 = 2h + 2w, h = 2w − 1
58 = 2(2w - 1) + 2w,
58 = 4w - 2 + 2w,
60 = 6w,
w = 10
h = 2(10) - 1, h = 19
Height is 19 cm, width is 10 cm.

59. Let q be the number of quarters and n be the number of nickels.
0.25q + 0.05n = 4.60, q + n = 24
q = 24 − n
0.25(24 - n) + 0.05n = 4.60
6 - 0.25n + 0.05n = 4.60
6 = 4.60 + 0.20n
1.40 = 0.20n
n = 7
q = 24 - 7, q = 17
Yoko has 17 quarters and 7 nickels.

61. Let d be the number of dimes and q be the number of quarters.
d + q = 28, 0.10d + 0.25q = 4.45
d = 28 - q,
0.10(28 - q) + 0.25q = 4.45,
2.80 - 0.10q + 0.25q = 4.45,
0.15q = 1.65,
q = 11
d = 28 - 11, d = 17
Chen Chen has 17 dimes and 11 quarters.

63. Let p = grams of peanuts and c = grams of cashews.
p + c = 270, 0.27p + 0.16c = 66
c = 270 - p,
0.27p + 0.16(270 - p) = 66
0.27p + 43.2 - 0.16p = 66
0.11p = 22.8
p ≈ 207
c = 270 - 207, c = 63
Mixture contains approximately 207 g of peanuts and 63 g of cashews.

65. a. A + B = 180, A = B + 24
(B + 24) + B = 180,
2B = 156,
B = 78
A = 78 + 24, A = 102
Angle A is 102° and angle B is 78°.

b. C + D = 180, D = C + 26
C + (C + 26) = 180,
2C = 154,
C = 77
D = 77 + 26, D = 103
Angle C is 77° and angle D is 103°.

c. E + F = 90, F = E + 2
E + (E + 2) = 90,
2E = 88,
E = 44
F = 44 + 2, F = 46
Angle E is 44° and angle F is 46°.

d. $G + H = 90$, $H = 2G + 3$
$G + (2G + 3) = 90$,
$3G = 87$,
$G = 29$
$H = 2(29) + 3$, $H = 61$
Angle G is 29° and angle H is 61°.

e. $I + J = 180$, $J = 2I - 45$,
$I + (2I - 45) = 180$,
$3I = 225$,
$I = 75$
$J = 2(75) - 45$, $J = 105$
Angle I is 75° and angle J is 105°.

67. Replace one variable with its' equivalent equation in terms of the other variable. Solve for that variable and use its value to find the other variable.

69. Choose guess and check for the simplest systems and graphing for the more complicated systems. Substitution can be used for all systems and is generally used to check answers.

Mid-Chapter 7 Test

1. $2x - y = 5000$,
$2x = 5000 + y$,
$y = 2x - 5000$

2. $V = \dfrac{\pi r^2 h}{3}$,
$3V = \pi r^2 h$,
$h = \dfrac{3V}{\pi r^2}$

3. $\dfrac{3}{2}x + \dfrac{2}{3}y = \dfrac{1}{4}$
$12\left(\dfrac{3}{2}x + \dfrac{2}{3}y\right) = 12 \cdot \dfrac{1}{4}$
$18x + 8y = 3$
$8y = -18x + 3$
$y = -\dfrac{18}{8}x + \dfrac{3}{8}$
$y = -\dfrac{9}{4}x + \dfrac{3}{8}$

4. $x + y = 5000$, $3x - 2y = -2500$,
$y = 5000 - x$,
$3x - 2(5000 - x) = -2500$,
$3x - 10,000 + 2x = -2500$,
$5x = 7500$,
$x = 1500$
$y = 5000 - 1500$, $y = 3500$

5. $x + y = 5000$, $3x - 2y = 2500$,
$y = 5000 - x$,
$3x - 2(5000 - x) = 2500$,
$3x - 10,000 + 2x = 2500$,
$5x = 12,500$,
$x = 2500$,
$y = 5000 - 2500$, $y = 2500$

6. a. It is the intersection and therefore the solution of $y = 6 - \frac{2}{3}x$ and $y = x + 1$.

b. Estimates should be between 4 and 5 for x and between 5 and 6 for y.

c. $y = 10 - x$, $y = x + 1$
$10 - x = x + 1$,
$10 = 2x + 1$,
$9 = 2x$,
$x = 4.5$
$y = 4.5 + 1$, $y = 5.5$
Intersection is at (4.5, 5.5).

d. $y = 10 - x$, $y = 6 - \frac{2}{3}x$
$10 - x = 6 - \frac{2}{3}x$,
$10 = 6 + \frac{1}{3}x$,
$4 = \frac{1}{3}x$,
$x = 12$
$y = 10 - 12$, $y = -2$
Intersection is at (12, -2).

7. a.

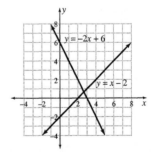

b. Estimates should be between 2 and 3 for x and between 0 and 1 for y.

c. $2x + y = 6,\ y = x - 2$
$2x + (x - 2) = 6,$
$3x - 2 = 6,$
$3x = 8,$
$x = \frac{8}{3}$
$y = \frac{8}{3} - 2,\ \ y = \frac{8-6}{3},\ \ y = \frac{2}{3}$
Intersection is at $\left(\frac{8}{3}, \frac{2}{3}\right)$.

8. $2x + y = 4,\ y = 4 - 2x,$
$2x + (4 - 2x) = 4,$
$2x + 4 - 2x = 4,$
$4 = 4,$ Always true, the system of equations has an infinite number of solutions; thus the lines coincident.

9. Let x = turtle eggs and y = ostrich eggs,
$x = 12y,\ x + y = 195$
$12y + y = 195,$
$13y = 195,$
$y = 15$
$x = 12(15),\ x = 180$
180 green turtle eggs and 15 ostrich eggs

10. Let x = 5 pound bags and y = 10 pound bags.
$x = 3y,\ 5x + 10y = 10,000$
$5(3y) + 10y = 10,000,$
$15y + 10y = 10,000,$
$25y = 10,000,$
$y = 400$
$x = 3(400),\ x = 1200$
1200 5-pound bags and 400 10-pound bags.

11. The ordered pair for the point of intersection is where the same input gives the same output in each equation.

Section 7.4

1. a.

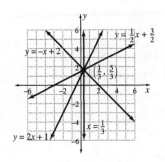

b.
$$\begin{array}{r} y = 2x + 1 \\ +\quad y = -x + 2 \\ \hline 2y = x + 3 \end{array}$$
$$y = \frac{x}{2} + \frac{3}{2}$$
Equation is on graph above.

c.
$$\begin{array}{r} y = 2x + 1 \\ -\quad y = -x + 2 \\ \hline 0 = 3x - 1 \end{array}$$
$$3x = 1$$
$$x = \frac{1}{3},$$
Equation is on graph above.

d. All graphs intersect at the same point $\left(\frac{1}{3}, 1\frac{2}{3}\right)$

3. $\begin{array}{r} x + y = -2 \\ x - y = 8 \\ \hline 2x = 6 \\ x = 3 \end{array}$
$x + y = -2$
$3 + y = -2$
$y = -5$
The solution is $x = 3, y = -5$.

5. $\begin{array}{r} m + n = 3 \\ -m + n = -11 \\ \hline 2n = -8 \\ n = -4 \end{array}$
$m + n = 3$
$m + (-4) = 3$
$m = 7$
The solution is $m = 7, n = -4$.

7. $\begin{array}{l} 2x + y = -1 \\ x + 2y = 4 \end{array}$ $\times(-2)$ $\begin{array}{r} 2x + y = -1 \\ -2x - 4y = -8 \\ \hline -3y = -9 \\ y = 3 \end{array}$
$2x + y = -1$
$2x + 3 = -1$
$2x = -4$
$x = -2$
The solution is $x = -2, y = 3$.

9. $2a+b=-5$ $2a+b=-5$

$a+3b=35$ $\times(-2)$ $-2a-6b=-70$

$-5b=-75$

$b=15$

$2a+b=-5$

$2a+15=-5$

$2a=-20$

$a=-10$

The solution is $a=-10, b=15$.

11. $2x+3y=3$ $\times(3)$ $6x+9y=9$

$3x-4y=-21$ $\times(-2)$ $-6x+8y=42$

$17y=51$

$y=3$

$2x+3y=3$

$2x+3(3)=3$

$2x+9=3$

$2x=-6$

$x=-3$

The solution is $x=-3, y=3$.

13. $5p-2q=-6$ $\times3$ $15p-6q=-18$

$2p+3q=9$ $\times2$ $4p+6q=18$

$19p=0$

$p=0$

$5p-2q=-6$

$5(0)-2q=-6$

$-2q=-6$

$q=3$

The solution is $p=0, q=3$.

15. $x+y=6$ $x+y=6$

$x+y=10$ $\times(-1)$ $-x-y=-10$

$0=-4$

A contradiction results since $0 \neq -4$. Thus, the system has no solution.

17. $x-y=7$ $\times(-2)$ $-2x+2y=-14$

$2x-2y=14$ $2x-2y=14$

$0=0$

An identity results since $0=0$ is always true. Thus, the system has an infinite number of solutions.

19. $x+y=5$ $x+y=5$

$y-x=-13$ $-x+y=-13$

$2y=-8$

$y=-4$

$x+y=5$

$x+(-4)=5$

$x=9$

The solution is $x=9, y=-4$.

21. $2x+3y=0$ $\times(3)$ $6x+9y=0$

$3x+2y=5$ $\times(-2)$ $-6x-4y=-10$

$5y=-10$

$y=-2$

$2x+3y=0$

$2x+3(-2)=0$

$2x-6=0$

$2x=6$

$x=3$

The solution is $x=3, y=-2$.

23. $7=5m+b$ $7=5m+b$

$3=3m+b$ $\times(-1)$ $-3=-3m-b$

$4=2m$

$2=m$

$7=5m+b$

$7=5(2)+b$

$7=10+b$

$-3=b$

The solution is $b=-3, m=2$.

25. $0.2x+0.6y=2.2$ $0.2x+0.6y=2.2$

$0.4x-0.2y=1.6$ $\times3$ $1.2x-0.6y=4.8$

$1.4x=7$

$x=5$

$0.2x+0.6y=2.2$

$0.2(5)+0.6y=2.2$

$1+0.6y=2.2$

$0.6y=1.2$

$y=2$

The solution is $x=5, y=2$.

27. $\begin{aligned} 0.5x + 0.2y &= 1.8 \;\times 3 \\ 0.2x - 0.3y &= -0.8 \;\times 2 \end{aligned}$ $\begin{aligned} 1.5x + 0.6y &= 5.4 \\ \underline{0.4x - 0.6y = -1.6} \\ 1.9x = 3.8 \\ x = 2 \end{aligned}$

$0.5x + 0.2y = 1.8$

$0.5(2) + 0.2y = 1.8$

$1 + 0.2y = 1.8$

$0.2y = 0.8$

$y = 4$

The solution is $x = 2, y = 4$.

29. $\begin{aligned} 3p + 4f &= 48 \;\times(-4) \\ 4p + 3f &= 43 \;\times 3 \end{aligned}$ $\begin{aligned} -12p - 16f &= -192 \\ \underline{12p + 9f = 129} \\ -7f = -63 \\ f = 9 \end{aligned}$

$3p + 4f = 48$

$3p + 4(9) = 48$

$3p + 36 = 48$

$3p = 12$

$p = 4$

The solution is $f = 9, p = 4$.

31. $x + y = 243$

$y = x - 17$

$x + (x - 17) = 243$

$x + x - 17 = 243$

$2x - 17 = 243$

$2x = 260$

$x = 130$

$y = x - 17$

$\quad = 130 - 17$

$\quad = 113$

The solution is $x = 130, y = 113$.

33. a. Let x represent one angle, and let y represent the other angle.

$\begin{aligned} x + y &= 90 \\ y &= x + 50 \end{aligned}$ $\begin{aligned} x + y &= 90 \\ \underline{-x + y = 50} \\ 2y = 140 \\ y = 70 \end{aligned}$

$x + y = 90$

$x + (70) = 90$

$x = 20$

The two angles measure $20°$ and $70°$.

b. Let x represent one angle, and let y represent the other angle.

$\begin{aligned} x + y &= 180 \\ y &= x + 50 \end{aligned}$ $\begin{aligned} x + y &= 180 \\ \underline{x - y = -50} \\ 2x = 130 \\ x = 65 \end{aligned}$

$y = x + 50$

$y = 65 + 50$

$y = 115$

The two angles measure $65°$ and $115°$.

35. a. $\begin{aligned} A + B &= 90 \\ A &= B - 22 \end{aligned}$ $\begin{aligned} A + B &= 90 \\ \underline{A - B = -22} \\ 2A = 68 \\ A = 34 \end{aligned}$

$A + B = 90$

$(34) + B = 90$

$B = 56$

The angles are $A = 34°$ and $B = 56°$.

b. $\begin{aligned} C + D &= 180 \\ D &= 4C + 5 \end{aligned}$ $\begin{aligned} C + D &= 180 \\ \underline{4C - D = -5} \\ 5C = 175 \\ C = 35 \end{aligned}$

$D = 4C + 5$

$\quad = 4(35) + 5$

$\quad = 140 + 5$

$\quad = 145$

The angles are $C = 35°$ and $D = 145°$.

c. $\begin{aligned} E + F &= 90 \\ E &= 4F + 10 \end{aligned}$ $\begin{aligned} E + F &= 90 \\ \underline{-E + 4F = -10} \\ 5F = 80 \\ F = 16 \end{aligned}$

$E = 4F + 10$

$\quad = 4(16) + 10$

$\quad = 64 + 10$

$\quad = 74$

The angles are $E = 74°$ and $F = 16°$.

131

d. $G + H = 180$ $G + H = 180$
     $G = 2H - 15$ $-G + 2H = 15$
$$3H = 195$$
$$H = 65$$
$$G + H = 180$$
$$G + (65) = 180$$
$$G = 115$$
The angles are $G = 115°$ and $H = 65°$.

e. $I + J = 180$ $I + J = 180$
     $J = 3I + 20$ $3I - J = -20$
$$4I = 160$$
$$I = 40$$
$$I + J = 180$$
$$(40) + J = 180$$
$$J = 140$$
The angles are $I = 40°$ and $J = 140°$.

37. Let x represent the larger number, and let y represent the smaller number.
$$x + y = 25$$
$$\underline{x - y = 8}$$
$$2x = 33$$
$$x = 16.5$$
$$x + y = 25$$
$$16.5 + y = 25$$
$$y = 8.5$$
The two numbers are 16.5 and 8.5.

39. Let x be the larger number, and let y be the smaller number.
    $x + y = 20$ $\times 2$ $2x + 2y = 40$
$2x - 2y = 21$ $\underline{2x - 2y = 21}$
$$4x = 61$$
$$x = 15.25$$
$$x + y = 20$$
$$15.25 + y = 20$$
$$y = 4.75$$
The two numbers are 15.25 and 4.75.

41. Let x be the cost of an adult ticket and y be the cost of a student ticket.
$$6x + 3y = 58.50 \quad \times 4$$
$$\underline{5x + 4y = 54 \quad \times(-3)}$$

$$24x + 12y = 234$$
$$\underline{-15x - 12y = -162}$$
$$9x = 72$$
$$x = 8$$
$$5x + 4y = 54$$
$$5(8) + 4y = 54$$
$$40 + 4y = 54$$
$$4y = 14$$
$$y = 3.5$$
An adult ticket costs $8.00 and a student ticket costs $3.50.

43. Let x be the price of a shirt and let y be the price of a tie.
$$3x + 2y = 109.95$$
$$4x + y = 119.95$$
Solve the second equation for y.
$$4x + y = 119.95$$
$$y = -4x + 119.95$$
Substitute the result into the first equation.
$$3x + 2y = 109.95$$
$$3x + 2(-4x + 119.95) = 109.95$$
$$3x - 8x + 239.90 = 109.95$$
$$-5x = -129.95$$
$$x = 25.99$$
$$y = -4x + 119.95$$
$$y = -4(25.99) + 119.95$$
$$y = 15.99$$
Each shirt costs $25.99 and each tie costs $15.99.

45. Let x be the cost of each CD and let y be the cost of each game.
$$2x + 3y = 137.95$$
$$4x + y = 75.95$$
Solve the second equation for y.
$$y = -4x + 75.95$$
Substitute this result into the first equation.

$$2x + 3(-4x + 75.95) = 137.95$$
$$2x - 12x + 227.85 = 137.95$$
$$-10x = -89.90$$
$$x = 8.99$$
$$y = -4x + 75.95$$
$$y = -4(8.99) + 75.95$$
$$y = -35.96 + 75.95$$
$$y = 39.99$$

Each CD is $8.99 and each game is $39.99.

47. Let x be the calories in a sugar cookie and y be the calories in a ginger snap.

$$4x + 2y = 296 \quad \times 5$$
$$3x + 10y = 329 \quad \times -1$$

$$20x + 10y = 1480$$
$$-3x - 10y = -329$$
$$\overline{17x = 1151}$$
$$x \approx 67.7$$

$$4x + 2y = 296$$
$$4(67.7) + 2y \approx 296$$
$$270.8 + 2y \approx 296$$
$$2y \approx 25.2$$
$$y \approx 12.6$$

Each sugar cookie has about 67.7 calories and each ginger snap has about 12.6 calories.

49. Let x be the number of calories in a cherry and let y be the number of calories in a grape.

$$15x + 22y = 126 \quad \times 4$$
$$20x + 11y = 113 \quad \times(-3)$$

$$60x + 88y = 504$$
$$-60x - 33y = -339$$
$$\overline{55y = 165}$$
$$y = 3$$

$$15x + 22y = 126$$
$$15x + 22(3) = 126$$
$$15x + 66 = 126$$
$$15x = 60$$
$$x = 4$$

Each cherry has 4 calories and each graph has 3 calories.

51. Let x be the number of calories in a green olive and let y be the number of calories in a ripe olive.

$$8x + 5y = 285 \quad \times 2$$
$$4x + 10y = 330 \quad \times(-1)$$

$$16x + 10y = 570$$
$$-4x - 10y = -330$$
$$\overline{12x = 240}$$
$$x = 20$$

$$8x + 5y = 285$$
$$8(20) + 5y = 285$$
$$160 + 5y = 285$$
$$5y = 125$$
$$y = 25$$

Each green olive has 20 calories and each ripe olive has 25 calories.

53. Let x be the number of hours Ned drives at 40 mph and y be the number of hours he drives at 65 mph.

	Rate	Time	Distance
Gravel	40	x	$40x$
Paved	65	y	$65y$
Total		7	405

$$x + y = 7$$
$$40x + 65y = 405$$

Solve the first equation for y.
$$x + y = 7$$
$$y = -x + 7$$

Substitute this result into the second equation.
$$40x + 65(-x + 7) = 405$$
$$40x - 65x + 455 = 405$$
$$-25x = -50$$
$$x = 2$$

$$y = -x + 7$$
$$y = -2 + 7$$
$$y = 5$$

Ned drives 2 hours on the gravel road and 5 hours on the paved road.

55. Let r be the speed of the airplanes and let w be the speed of the wind.

	Net rate	Time	Distance
South	$r + w$	4	1100
North	$r - w$	5	1100

$$4(r+w)=1100 \div 4 \quad r+w=275$$
$$5(r-w)=1100 \div 5 \quad \underline{r-w=220}$$
$$2r=495$$
$$r=247.5$$

$$r+w=275$$
$$247.5+w=275$$
$$w=27.5$$

The airspeed of the planes is 247.5 mph and the wind speed is 27.5 mph.

57. Let r be the speed of the airplane and let w be the speed of the wind.

	Net rate	Time	Distance
With wind	$r + w$	2	950
Against wind	$r - w$	3	975

$$2(r+w)=950 \div 2 \quad r+w=475$$
$$3(r-w)=975 \div 3 \quad \underline{r-w=325}$$
$$2r=800$$
$$r=400$$

$$r+w=475$$
$$400+w=475$$
$$w=75$$

The airspeed of the plane is 400 mph and the wind speed is 75 mph.

59. Let r be the speed of the boat and let c be the speed of the current.

	Net rate	Time	Distance
With current	$r + c$	2	20
Against current	$r - c$	5	20

$$2(r+c)=20 \div 2 \quad r+c=10$$
$$5(r-c)=20 \div 5 \quad \underline{r-c=4}$$
$$2r=14$$
$$r=7$$

$$r+c=10$$
$$7+c=10$$
$$c=3$$

The speed of the boat (in still water) is 7 mph and the speed of the current is 3 mph.

61. It is possible to make forward progress when walking because of friction with the ground.

63. Multiply each equation by an appropriate constant (if necessary) to get coefficients on one variable that have the same absolute value, but different signs. Add the equations and solve for the remaining variable. Then use substitution to find the value of the other variable.

65. Multiply the first equation by e and the second equation by $-b$. Then add the two equations.

Section 7.5

1. $x+z=2$
$x-y=1$
$y-z=-2$

Solve the first equation for z.
$z=-x+2$
Solve the second equation for y.
$y=x-1$
Substitute the results into the third equation and solve for x.
$$y-z=-2$$
$$(x-1)-(-x+2)=-2$$
$$x-1+x-2=-2$$
$$2x-3=-2$$
$$2x=1$$
$$x=\frac{1}{2}$$

Substitute $x=\frac{1}{2}$ into the equations above for y and z.

$z=-x+2$	$y=x-1$
$z=-\dfrac{1}{2}+2$	$y=\dfrac{1}{2}-1$
$z=\dfrac{3}{2}$	$y=-\dfrac{1}{2}$

The solution to the system is $\left(\frac{1}{2},-\frac{1}{2},\frac{3}{2}\right)$ or $(0.5,-0.5,1.5)$.

3. $x + y = 2$

$z - x = 7$

$z - y = 2$

Solve the first equation for y.

$y = -x + 2$

Solve the second equation for z.

$z = x + 7$

Substitute these results into the third equation and solve for x.

$$z - y = 2$$
$$(x+7) - (-x+2) = 2$$
$$x + 7 + x - 2 = 2$$
$$2x + 5 = 2$$
$$2x = -3$$
$$x = -\frac{3}{2}$$

Substitute $x = -\frac{3}{2}$ into the equations above for y and z.

$y = -x + 2$ $z = x + 7$

$y = -\left(-\dfrac{3}{2}\right) + 2$ $z = -\dfrac{3}{2} + 7$

$y = \dfrac{7}{2}$ $z = \dfrac{11}{2}$

The solution to the system is $\left(-\frac{3}{2}, \frac{7}{2}, \frac{11}{2}\right)$ or $\left(-1\frac{1}{2}, 3\frac{1}{2}, 5\frac{1}{2}\right)$.

5. $x + z = 3$

$y + x = 4$

$y - z = 1$

Solve the first equation for z.

$z = -x + 3$

Solve the second equation for y.

$y = -x + 4$

Substitute these results into the third equation and solve for x.

$$y - z = 1$$
$$(-x+4) - (-x+3) = 1$$
$$-x + 4 + x - 3 = 1$$
$$1 = 1$$

This results in an identity because $1 = 1$ is always true. Thus, the system has an infinite number of solutions.

7. $x - y = 3$

$y + z = 4$

$x + z = 5$

Solve the first equation for y.

$y = x - 3$

Solve the third equation for z.

$z = -x + 5$

Substitute these results into the second equation and solve for x.

$$y + z = 4$$
$$(x-3) + (-x+5) = 4$$
$$x - 3 - x + 2 = 4$$
$$-1 = 4 \text{ false}$$

This results in a contradiction since $-1 \neq 4$. Thus, the system has no solution.

9. $x + y = 5$

$y + z = 5$

$z + x = 5$

Solve the first equation for y.

$y = -x + 5$

Solve the third equation for z.

$z = -x + 5$

Substitute these results into the second equation and solve for x.

$$y + z = 5$$
$$(-x+5) + (-x+5) = 5$$
$$-2x + 10 = 5$$
$$-2x = -5$$
$$x = \frac{5}{2}$$

Substitute $x = \frac{5}{2}$ into the equations above for y and z.

$y = -x + 5$ $z = -x + 5$

$y = -\left(\dfrac{5}{2}\right) + 5$ $z = -\left(\dfrac{5}{2}\right) + 5$

$y = \dfrac{5}{2}$ $z = \dfrac{5}{2}$

The solution to the system is $\left(\frac{5}{2}, \frac{5}{2}, \frac{5}{2}\right)$ or $(2.5, 2.5, 2.5)$.

11. a.   Answers will vary. One possibility is $(1, -3, 2)$

     b.   Answers will vary. Some possible solutions are $(1, 3, 2)$ and $(4, 0, -1)$.

c. Answers will vary. Some possible solutions are $(10, 2, -4)$ and $(7, -1, -1)$.

13. a. 1 equation

    b. 2 equations

    c. 3 equations

15. Let m be the amount Katreen earns mowing lawns, s be the amount she earns shopping for the elderly, and p be the amount she earns scraping old paint.
$$m + p + s = 1375$$
$$m = 2s$$
$$p = m + 500$$
Solve the second equation for s.
$$s = \frac{m}{2}$$
Substitute this result, and $p = m + 500$, into the first equation and solve for m.
$$m + p + s = 1375$$
$$m + (m + 500) + \frac{m}{2} = 1375$$
$$\frac{5}{2}m + 500 = 1375$$
$$\frac{5}{2}m = 875$$
$$m = 350$$
Substitute $m = 350$ into the equations for p and s above.
$$s = \frac{m}{2} = \frac{350}{2} = 175$$
$$p = m + 500 = 350 + 500 = 850$$
Katreen earns $350 mowing lawns, $175 shopping for the elderly, and $850 scraping old paint.

17. Let x be the number of 1-pound bags, y be the number of 2-pound bags, and z be the number of 5-pound bags.
$$x + 2y + 5z = 2400$$
$$x = 2y$$
$$x = 10z$$
Solve the second equation for y.
$$y = \frac{x}{2}$$
Solve the third equation for z.
$$z = \frac{x}{10}$$

Substitute these results into the first equation and solve for x.
$$x + 2\left(\frac{x}{2}\right) + 5\left(\frac{x}{10}\right) = 2400$$
$$x + x + \frac{x}{2} = 2400$$
$$\frac{5}{2}x = 2400$$
$$x = 960$$
Substitute $x = 960$ into the equations for y and z above.
$$y = \frac{x}{2} = \frac{960}{2} = 480$$
$$z = \frac{x}{10} = \frac{960}{10} = 96$$
The I. R. Rabbit Company should fill 960 1-pound bags, 480 2-pound bags, and 96 5-pound bags.

19. $A + B + C = 180$
$$A = 2B$$
$$B = 3C$$
Solve the third equation for C.
$$C = \frac{B}{3}$$
Substitute this result and $A = 2B$ into the first equation and solve for B.
$$A + B + C = 180$$
$$2B + B + \frac{B}{3} = 180$$
$$\frac{10}{3}B = 180$$
$$B = 54$$
Substitute $B = 54$ into the equations for A and C above.
$$A = 2B = 2(54) = 108$$
$$C = \frac{B}{3} = \frac{54}{3} = 18$$
The solution of the system is $A = 108, B = 54$, and $C = 18$.

21. $A + B + C = 180$
$$A = 2B$$
$$C = A + 20$$
Solve the second equation for B.
$$B = \frac{A}{2}$$
Substitute this result and $C = A + 20$ into the

first equation and solve for *A*.

$$A + B + C = 180$$

$$A + \frac{A}{2} + (A + 20) = 180$$

$$\frac{5}{2}A + 20 = 180$$

$$\frac{5}{2}A = 160$$

$$A = 64$$

Substitute $A = 64$ into the equations for *B* and *C* above.

$$B = \frac{A}{2} = \frac{64}{2} = 32$$

$$C = A + 20 = 64 + 20 = 84$$

The angle measures are $A = 64°$, $B = 32°$, and $C = 84°$.

23. Let *A* and *B* represent the measures of the two equal angles, and *C* represent the measure of the third angle.

$$A = B$$

$$A + B + C = 180$$

$$A = \frac{1}{2}C$$

Substitute $B = A$ and $A = \frac{1}{2}C$ into the second equation and solve for *C*.

$$A + B + C = 180$$

$$\frac{1}{2}C + \frac{1}{2}C + C = 180$$

$$2C = 180$$

$$C = 90$$

Substitute $C = 90$ into the equations for *A* and *B* above.

$$A = \frac{1}{2}C = \frac{1}{2}(90) = 45$$

$$B = A = 45$$

The angle measures are $A = B = 45°$ and $C = 90°$.

25. Let *l* represent the length, let *w* represent the width, and let *h* represent the height of the box.

$$l = w + 7$$

$$w = 4h$$

$$l + w + h = 34$$

Solve the second equation for *h*.

$$h = \frac{w}{4}$$

Substitute this result and $l = w + 7$ into the third

equation and solve for *w*.

$$l + w + h = 34$$

$$(w + 7) + w + \frac{w}{4} = 34$$

$$\frac{9}{4}w + 7 = 34$$

$$\frac{9}{4}w = 27$$

$$w = 12$$

Substitute $w = 12$ into the equations for *l* and *h* above.

$$l = w + 7 = 12 + 7 = 19$$

$$h = \frac{w}{4} = \frac{12}{4} = 3$$

The length of the box is 19 inches, the width of the box is 12 inches, and the height of the box is 3 inches.

Section 7.6

1. a. $0 \le 0 + 2$, true
      $0 \ge 0 - 2$, true
      (0, 0) is a solution

    b. $1 \le 1 + 2$, true
      $1 \ge 1 - 2$, true
      (1, 1) is a solution

    c. $-2 \le 1 + 2$, true
      $-2 \ge 1 - 2$, false
      (1, -2) is not a solution

    d. $3 \le -3 + 2$, false
      (-3, 3) is not a solution

3. a. $0 + 0 > 4$, false
      (0, 0) is not a solution

    b. $2 + 2 > 4$, false
      (2, 2) is not a solution

    c. $5 + 0 > 4$, true
      $5 - 0 > 4$, true
      (5, 0) is a solution

    d. $0 + 5 > 4$, true
      $0 - 5 > 4$, false
      (0, 5) is not a solution

5. Graph the inequality $y \le 2x + 2$. Boundary line: $y = 2x + 2$. Because the inequality contains $\le$, use a solid line.

Test Point: $(0,0)$: $0 \le 2(0)+2$ is true.

Shade the side containing $(0,0)$.

Graph the inequality $y \ge -x+4$. Boundary line: $y = -x+4$. Because the inequality contains $\ge$, use a solid line.

Test Point: $(0,0)$: $0 \ge -(0)+4$ is false.

Shade the side not containing $(0,0)$.

The overlapping shaded region (that is, the shaded region in the graph below) is the solution to the system of inequalities.

The corner point is $\left(\frac{2}{3}, \frac{10}{3}\right)$.

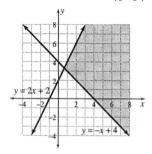

7. Graph the inequality $y \le 2x+2$. Boundary line: $y = 2x+2$. Because the inequality contains $\le$, use a solid line.

    Test Point: $(0,0)$: $0 \le 2(0)+2$ is true. Shade the side containing $(0,0)$.

    Graph the inequality $y \ge 3x-3$. Boundary line: $y = 3x-3$. Because the inequality contains $\ge$, use a solid line.

    Test Point: $(0,0)$: $0 \ge 3(0)-3$ is true. Shade the side containing $(0,0)$.

    The overlapping shaded region (that is, the shaded region in the graph below) is the solution to the system of inequalities.

    The corner point is $(5,12)$.

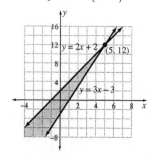

9. Graph the inequality $y \ge 2x-3$. Boundary line: $y = 2x-3$. Because the inequality contains $\ge$, use a solid line.
    Test Point: $(0,0)$: $0 \ge 2(0)-3$ is true.

    Shade the side containing $(0,0)$.

    Graph the inequality $y > -x+3$. Boundary line: $y = -x+3$. Because the inequality contains $>$, use a dashed line.

    Test Point: $(0,0)$: $0 > -(0)+3$ is false.

    Shade the side not containing $(0,0)$.

    The overlapping shaded region (that is, the shaded region in the graph below) is the solution to the system of inequalities.

    The corner point is $(2,1)$.

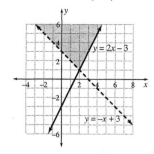

11. Graph the inequality $y > 3x-3$. Boundary line: $y = 3x-3$. Because the inequality contains $>$, use a dashed line.
    Test Point: $(0,0)$: $0 > 3(0)-3$ is true.

    Shade the side containing $(0,0)$.

    Graph the inequality $y < 2x-2$. Boundary line: $y = 2x-2$. Because the inequality contains $<$, use a dashed line.

    Test Point: $(0,0)$: $0 < 2(0)-2$ is false.

    Shade the side not containing $(0,0)$.

    The overlapping shaded region (that is, the shaded region in the graph below) is the solution to the system of inequalities.

The corner point is $(1,0)$.

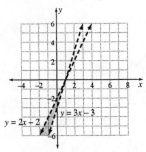

13. Graph the inequality $x \geq 2$.
Boundary line: $x = 2$. Because the inequality contains $\geq$, use a solid line.
Test Point: $(0,0)$: $0 \geq 2$ is false. Shade the side not containing $(0,0)$.

Graph the inequality $y \geq 3$.
Boundary line: $y = 3$. Because the inequality contains $\geq$, use a solid line.
Test Point: $(0,0)$: $0 \geq 3$ is false. Shade the side not containing $(0,0)$.

The overlapping shaded region (that is, the shaded region in the graph below) is the solution to the system of inequalities.

The corner point is $(2,3)$.

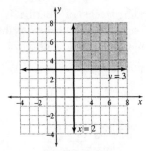

15. Graph the inequality $x > -2$.
Boundary line: $x = -2$. Because the inequality contains $>$, use a dashed line.
Test Point: $(0,0)$: $0 > -2$ is true. Shade the side containing $(0,0)$.

Graph the inequality $y \leq 3$.
Boundary line: $y = 3$. Because the inequality contains $\leq$, use a solid line.
Test Point: $(0,0)$: $0 \leq 3$ is true. Shade the side containing $(0,0)$.

The overlapping shaded region (that is, the shaded region in the graph below) is the solution to the system of inequalities.

The corner point is $(-2,3)$.

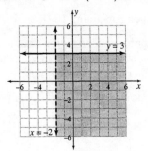

17. Graph the inequality $x > -2$.
Boundary line: $x = -2$. Because the inequality contains $>$, use a dashed line.
Test Point: $(0,0)$: $0 > -2$ is true. Shade the side containing $(0,0)$.

Graph the inequality $y < -1$.
Boundary line: $y = -1$. Because the inequality contains $<$, use a dashed line.
Test Point: $(0,0)$: $0 < -1$ is false. Shade the side not containing $(0,0)$.

The overlapping shaded region (that is, the shaded region in the graph below) is the solution to the system of inequalities.

The corner point is $(-2,-1)$.

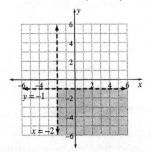

19. The two inequalities $x \geq 0$ and $y \geq 0$ require that the graph be in quadrant I or on the non-negative coordinate axes.

Graph the inequality $y \leq 4$.
Boundary line: $y = 4$. Because the inequality contains $\leq$, use a solid line.
Test Point: $(0,0)$: $0 \leq 4$ is true. Shade the side containing $(0,0)$.

Graph the inequality $y \le -2x + 5$.
Boundary line: $y = -2x + 5$. Because the
inequality contains $\le$, use a solid line.
Test Point: $(0,0)$: $0 \le -2(0) + 5$ is true.

Shade the side containing $(0,0)$.

The overlapping shaded region (that is, the
shaded region in the graph below) is the solution
to the system of linear inequalities.

The corner points are $(0,0)$, $\left(\frac{5}{2}, 0\right)$, $\left(\frac{1}{2}, 4\right)$, and
$(0,4)$.

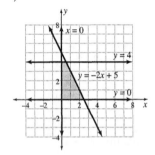

21. The two inequalities $x \ge 0$ and $y \ge 0$ require
that the graph be in quadrant I or on the non-
negative coordinate axes.

Graph the inequality $x + y \ge 3$.

Boundary line: $x + y = 3$ $(y = -x + 3)$. Because
the inequality contains $\ge$, use a solid line.
Test Point: $(0,0)$: $0 + 0 \ge 3$ is false.

Shade the side not containing $(0,0)$.

Graph the inequality $x + y \le 5$.

Boundary line: $x + y = 5$ $(y = -x + 5)$. Because
the inequality contains $\le$, use a solid line.
Test Point: $(0,0)$: $0 + 0 \le 5$ is true.

Shade the side containing $(0,0)$.

The overlapping shaded region (that is, the
shaded region in the graph below) is the solution
to the system of linear inequalities.

The corner points are $(0,3)$, $(3,0)$, $(5,0)$, and
$(0,5)$.

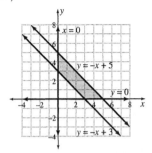

23. The two inequalities $x \le 0$ and $y \ge 0$ require
that the graph be in quadrant II or on the
coordinate axes bordering the quadrant.

Graph the inequality $y \le -x + 2$.

Boundary line: $y = -x + 2$. Because the
inequality contains $\le$, use a solid line.
Test Point: $(0,0)$: $0 \le -(0) + 2$ is true.

Shade the side containing $(0,0)$.

Graph the inequality $x \ge -3$.
Boundary line: $x = -3$. Because the inequality
contains $\ge$, use a solid line.
Test Point: $(0,0)$: $0 \ge -3$ is true. Shade the side

containing $(0,0)$.

The overlapping shaded region (that is, the
shaded region in the graph below) is the solution
to the system of linear inequalities.

The corner points are $(0,0)$, $(-3,0)$, $(-3,5)$,

and $(0,2)$.

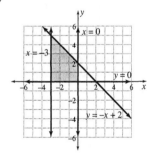

25. $x \ge 0$

$y \ge 0$

$x \le 4$

$x + y \le 12$

The two inequalities $x \ge 0$ and $y \ge 0$ require
that the graph be in quadrant I or on the non-

negative coordinate axes.
Graph the inequality $x + y \leq 12$.

Boundary line: $x + y = 12$ $(y = -x + 12)$.

Because the inequality contains $\leq$, use a solid line.

Test Point: $(0,0)$: $0 + 0 \leq 12$ is true. Shade the side containing $(0,0)$.

Graph the inequality $x \leq 4$.
Boundary line: $x = 4$. Because the inequality contains $\leq$, use a solid line.
Test Point: $(0,0)$: $0 \leq 4$ is true. Shade the side containing $(0,0)$.

The overlapping shaded region (that is, the shaded region in the graph below) is the solution to the system of linear inequalities.

The corner points are $(0,0)$, $(4,0)$, $(0,12)$ and $(4,8)$.

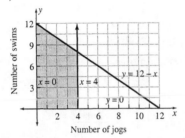

27. $x \geq 0$

$y \geq 0$

$y \leq 12$

$15x + 8y \leq 240$

The two inequalities $x \geq 0$ and $y \geq 0$ require that the graph be in quadrant I or on the non-negative coordinate axes.
Graph the inequality $15x + 8y \leq 240$.

Boundary line: $15x + 8y = 240$ $\left(y = -\frac{15}{8}x + 30\right)$.

Because the inequality contains $\leq$, use a solid line.

Test Point: $(0,0)$: $15(0) + 8(0) \leq 240$ is true.

Shade the side containing $(0,0)$.

Graph the inequality $y \leq 12$.
Boundary line: $y = 12$. Because the inequality contains $\leq$, use a solid line.
Test Point: $(0,0)$: $0 \leq 12$ is true. Shade the side containing $(0,0)$.

The overlapping shaded region (that is, the shaded region in the graph below) is the solution to the system of linear inequalities.

The corner points are $(0,0)$, $(0,12)$, $(16,0)$ and $\left(\frac{48}{5}, 12\right)$.

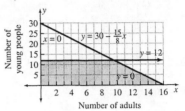

29. $x \geq 0$

$y \geq 0$

$y \leq 5000$

$x + y \leq 45,000$

The two inequalities $x \geq 0$ and $y \geq 0$ require that the graph be in quadrant I or on the non-negative coordinate axes.
Graph the inequality $x + y \leq 45,000$.

Boundary line: $x + y = 45,000$

$(y = -x + 45,000)$. Because the inequality contains $\leq$, use a solid line.

Test Point: $(0,0)$: $0 + 0 \leq 45,000$ is true. Shade the side containing $(0,0)$.

Graph the inequality $y \leq 5000$.
Boundary line: $y = 5000$. Because the inequality contains $\leq$, use a solid line.
Test Point: $(0,0)$: $0 \leq 5000$ is true. Shade the side containing $(0,0)$.

The overlapping shaded region (that is, the shaded region in the graph below) is the solution to the system of linear inequalities.

The corner points are $(0,0)$, $(45000,0)$, $(0,5000)$ and $(40000,5000)$.

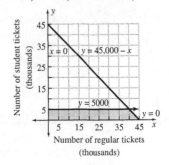

Chapter 7 Review Exercises

1. $y - 2x = 5000,$
$y = 5000 + 2x$

3. $C = 2\pi r,$
$$r = \frac{C}{2\pi}$$

5. $\dfrac{a}{b} = \dfrac{5}{8},$
$5b = 8a,$
$$b = \frac{8a}{5}$$

7.

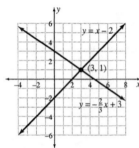

The solution is $(3,1)$.

9.

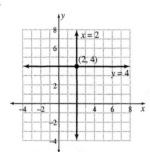

The solution is $(2,4)$.

11. $y = x + 2$
$y = 3x + 3$
Substitute $y = x + 2$ into the second equation and solve for x.
$$y = 3x + 3$$
$$(x+2) = 3x + 3$$
$$-1 = 2x$$
$$-\frac{1}{2} = x$$
Substitute $x = -\frac{1}{2}$ into the first equation and

solve for y.
$$y = x + 2 = -\frac{1}{2} + 2 = \frac{3}{2}$$
The solution is $x = -\frac{1}{2}, y = \frac{3}{2}$.

13. $x + y = 5000$
$3x - 2y = 12,500$
Solve the first equation for y.
$y = -x + 5000$
Substitute this result into the second equation and solve for x.
$$3x - 2y = 12,500$$
$$3x - 2(-x + 5000) = 12,500$$
$$3x + 2x - 10,000 = 12,500$$
$$5x = 22,500$$
$$x = 4500$$
Substitute $x = 4500$ into the equation for y above.
$$y = -x + 5000$$
$$y = -4500 + 5000$$
$$y = 500$$
The solution is $x = 4500, y = 500$.

15.
$$\begin{array}{ll} 3 = -2m + b & -2m + b = 3 \\ 2 = 3m + b & \underline{-3m - b = -2} \\ & -5m = 1 \\ & m = -\dfrac{1}{5} \end{array}$$
$$3 = -2m + b$$
$$3 = -2\left(-\frac{1}{5}\right) + b$$
$$3 = \frac{2}{5} + b$$
$$\frac{13}{5} = b$$
The solution is $m = -\frac{1}{5}, b = \frac{13}{5}$.

17.
$$\begin{array}{ll} 2x - 3y = 7 & 2x - 3y = 7 \\ 3y + 4x = -1 & \underline{4x + 3y = -1} \\ & 6x = 6 \\ & x = 1 \end{array}$$

$$2x - 3y = 7$$
$$2(1) - 3y = 7$$
$$2 - 3y = 7$$
$$-3y = 5$$
$$y = -\frac{5}{3}$$

The solution is $x = 1, y = -\frac{5}{3}$.

19. $\begin{array}{ll} 5x + 3y = -18 & \times 2 \\ 3x - 2y = -7 & \times 3 \end{array}$ $\begin{array}{l} 10x + 6y = -36 \\ \underline{9x - 6y = -21} \\ 19x = -57 \\ x = -3 \end{array}$

$$5x + 3y = -18$$
$$5(-3) + 3y = -18$$
$$-15 + 3y = -18$$
$$3y = -3$$
$$y = -1$$

The solution is $x = -3, y = -1$.

21. $x + y = 15$
$y = -6x$

Substitute $y = -6x$ into the first equation and solve for x.
$$x + y = 15$$
$$x + (-6x) = 15$$
$$-5x = 15$$
$$x = -3$$

Substitute $x = -3$ into the second equation and solve for y.
$$y = -6x = -6(-3) = 18$$

The solution is $x = -3, y = 18$.

23. $\begin{array}{llll} 3y - 4x = 2 & \rightarrow & 4x - 3y = -2 \\ 8x = 6y + 4 & \rightarrow & 8x - 6y = 4 \end{array}$

$\begin{array}{ll} 4x - 3y = -2 & \times(-2) \\ 8x - 6y = 4 \end{array}$ $\begin{array}{l} -8x + 6y = 4 \\ \underline{8x - 6y = 4} \\ 0 = 8 \end{array}$

This results in a contradiction since $0 \neq 8$. Thus, the system has no solution.

25. Since $0 = 0$ is always true (an identity), the lines are coincident.

27. Let x be the number of legs on the record centipede and y be the number of legs on the record millipede.
$$x = y - 356$$
$$x + y = 1064$$

Substitute $x = y - 356$ into the second equation and solve for x.
$$x + y = 1064$$
$$(y - 356) + y = 1064$$
$$2y - 356 = 1064$$
$$2y = 1420$$
$$y = 710$$

Substitute $y = 710$ into the equation for x above.
$$x = y - 356$$
$$x = 710 - 356$$
$$x = 354$$

The record centipede had 354 legs and the record millipede had 710 legs.

29. Let x be the number of calories in an English muffin and y be the number of calories in a fried egg.

$\begin{array}{ll} x + 2y = 330 \\ 3x + y = 515 & \times(-2) \end{array}$ $\begin{array}{l} x + 2y = 330 \\ \underline{-6x - 2y = -1030} \\ -5x = -700 \\ x = 140 \end{array}$

$$x + 2y = 330$$
$$140 + 2y = 330$$
$$2y = 190$$
$$y = 95$$

Each English muffin contains 140 calories and each fried egg contains 95 calories.

31. Let x be the number of cows and y be the number of ducks.

$\begin{array}{ll} x + y = 20 & \times(-2) \\ 4x + 2y = 64 \end{array}$ $\begin{array}{l} -2x - 2y = -40 \\ \underline{4x + 2y = 64} \\ 2x = 24 \\ x = 12 \end{array}$

$$x + y = 20$$
$$12 + y = 20$$
$$y = 8$$

Mr. McFadden has 12 cows and 8 ducks.

33. Let r represent the speed of a trout and c represent the speed of the current.

	Rate	Time	Distance
With current	$r+c$	0.8	16
Against current	$r-c$	0.8	8

$$0.8(r+c)=16 \quad \div 0.8 \quad r+c=20$$
$$\underline{0.8(r-c)=8 \quad \div 0.8 \quad r-c=10}$$
$$2r=30$$
$$r=15$$

$$r+c=20$$
$$15+c=20$$
$$c=5$$

The trout's speed in still water is 15 mph and the speed of the current is 5 mph.

35. Let x be the interest rate of the car loan and y be the interest rate of the credit card.
$$800x+2200y=452 \quad \div 800$$
$$\underline{2000x+1000y=320 \quad \div(-2000)}$$

$$x+2.75y=0.565$$
$$\underline{-x-0.5y=-0.16}$$
$$2.25y=0.405$$
$$y=0.18$$
$$x+2.75y=0.565$$
$$x+2.75(0.18)=0.565$$
$$x+0.495=0.565$$
$$x=0.07$$

The interest rate on the car loan is 7% and the interest rate on the credit card is 18%.

37. $A+B=90$

$A=2B+10$

Substitute $A=2B+10$ into the first equation and solve for B.
$$A+B=90$$
$$(2B+10)+B=90$$
$$3B+10=90$$
$$3B=80$$
$$B=\frac{80}{3}$$

Substitute $B=\frac{80}{3}$ into the second equation and solve for A.

$$A=2B+10$$
$$A=2\left(\frac{80}{3}\right)+10$$
$$A=\frac{160}{3}+10$$
$$A=\frac{190}{3}$$

The angle measures are $A=\frac{80}{3}°$ and $B=\frac{190}{3}°$ (or $26\frac{2}{3}°$ and $63\frac{1}{3}°$).

39. $A+B+C=180$

$2A=B$

$6A=C$

Substitute $B=2A$ and $C=6A$ into the first equation and solve for A.
$$A+B+C=180$$
$$A+2A+6A=180$$
$$9A=180$$
$$A=20$$

Substitute $A=20$ into the equations for B and C above.
$$B=2A=2(20)=40$$
$$C=6A=6(20)=120$$

The angle measures are $A=20°$, $B=40°$, and $C=120°$.

41. $x+y=6$

$y+z=8$

$x+z=7$

Solve the first equation for y.
$$y=-x+6$$

Solve the third equation for z.
$$z=-x+7$$

Substitute these results into the second equation and solve for x.
$$y+z=8$$
$$(-x+6)+(-x+7)=8$$
$$-2x+13=8$$
$$-2x=-5$$
$$x=\frac{5}{2}$$

Substitute $x=\frac{5}{2}$ into the equations for y and z above.

$$y = -x + 6 \qquad\qquad z = -x + 7$$

$$y = -\frac{5}{2} + 6 \qquad\qquad z = -\frac{5}{2} + 7$$

$$y = \frac{7}{2} \qquad\qquad z = \frac{9}{2}$$

The solution is $\left(\frac{5}{2}, \frac{7}{2}, \frac{9}{2}\right)$ or $(2.5, 3.5, 4.5)$.

43. $x + y = 6$

$y + z = 8$

$z - x = 2$

Solve the first equation for y.

$y = -x + 6$

Solve the third equation for z.

$z = x + 2$

Substitute these results into the second equation and solve for x.

$$y + z = 8$$

$$(-x + 6) + (x + 2) = 8$$

$$8 = 8$$

This results in an identity. Thus, there are an infinite number of solutions to the system.

45. Let x = one of the two equal sides and y = the third side.

$2x + y = 32$

$x = y + 2.5$

$2(y + 2.5) + y = 32,$

$2y + 5 + y = 32,$

$3y = 27,$

$y = 9$

$x = 9 + 2.5, x = 11.5$

The 2 equal sides are each 11.5 inches and the third side is 9 inches.

47. Let p = protein, f = fat and c = carbohydrates

4p + 9f + 4c = 198

p + f + c = 37

p = c + 5,

substitute for p in the other 2 equations

4(c + 5) + 9f + 4c = 198,

4c + 20 + 9f + 4c = 198

9f + 8c = 178   (eqn. A)

(c + 5) + f + c = 37,

f + 2c = 32, solve this for f

f = 32 - 2c,

substitute this for f in eqn. A

9(32 - 2c) + 8c = 178,

288 - 18c + 8c = 178,

-10c = -110,

c = 11,

use this to find f and p

f = 32 - 2(11), f = 10,

p = 11 + 5, p = 16,

There are 16 g of protein, 10 g of fat and 11 g of carbohydrate in this serving of shrimp.

49. Let p = protein, f = fat and
c = carbohydrates

4p + 9f + 4c = 903

p + f + c = 137

f = c + 44,

substitute for f in the other 2 equations.

4p + 9(c + 44) + 4c = 903,

4p + 9c + 396 + 4c = 903,

4p + 13c = 507   (eqn. B)

p + (c + 44) + c = 137,

p + 2c = 93,

solve this for p

p = 93 - 2c,

substitute this for p in eqn. B

4(93 - 2c) + 13c = 507,

372 - 8c + 13c = 507,

5c = 135,

c = 27,

use this to find p and f

p = 93 - 2(27), p = 39,

f = 27 + 44, f = 71

There are 39 g of protein, 71 g of fat and 27 g of carbohydrates in the peanuts.

51. Answers and explanations may vary.
Suggestions:

a. Choose substitution when one variable has a 1 or -1 coefficient.

b. Choose guess and check when you are trying to understand a problem.

c. Choose a table and graph from a graphing calculator when there is no algebraic method for a solution.

d. Choose elimination when coefficients are integers or with multiplication can be made into integers.

e. Choose graphing when the equations are easily placed into y = mx + b form.

53. $x + y < 4 \rightarrow y < -x + 4$

Graph the inequality $y < -x + 4$. Boundary line: $y = -x + 4$. Because the inequality contains $<$, use a dashed line.

Test Point: $(0,0)$: $0 < -(0) + 4$ is true.

Shade the side containing $(0,0)$.

Graph the inequality $y \geq x$. Boundary line: $y = x$. Because the inequality contains $\geq$, use a solid line.

Test Point: $(0,2)$: $2 \geq 0$ is true.

Shade the side containing $(0,0)$.

The overlapping shaded region (that is, the shaded region in the graph below) is the solution to the system of inequalities.

The corner point is $(2,2)$.

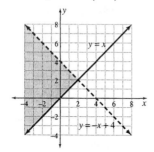

55. Graph the inequality $x \leq 2$.
Boundary line: $x = 2$. Because the inequality contains $\leq$, use a dashed line.
Test Point: $(0,0)$: $0 \leq 2$ is true.

Shade the side containing $(0,0)$.

Graph the inequality $y < 1$.
Boundary line: $y = 1$. Because the inequality contains $<$, use a dashed line.
Test Point: $(0,0)$: $0 < 1$ is true.

Shade the side containing $(0,0)$.

The overlapping shaded region (that is, the shaded region in the graph below) is the solution to the system of inequalities.

The corner point is $(2,1)$.

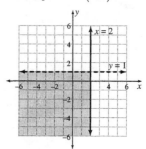

57. The two inequalities $x \geq 0$ and $y \geq 0$ require that the graph be in quadrant I or along the non-negative coordinate axes.

Graph the inequality $y \leq 4$.

Boundary line: $y = 4$. Because the inequality contains $\leq$, use a solid line.

Test Point: $(0,0)$: $0 \leq 4$ is true. Shade the side containing $(0,0)$.

Graph the inequality $y \leq -x + 6$.

Boundary line: $y = -x + 6$. Because the inequality contains $\leq$, use a solid line.

Test Point: $(0,0)$: $0 \leq -(0) + 6$ is true.

Shade the side containing $(0,0)$.

The overlapping shaded region (that is, the shaded region in the graph below) is the solution to the system of linear inequalities.

The corner points are $(0,0)$, $(6,0)$, $(2,4)$, and $(0,4)$.

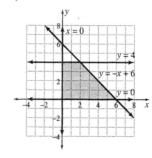

59. $x \geq 0$

$y \geq 0$

$y \leq 2$

$55x + 65y \leq 715$

The two inequalities $x \geq 0$ and $y \geq 0$ require that the graph be in quadrant I or along the non-negative coordinate axes.

Graph the inequality $55x + 65y \le 715$.
Boundary line: $55x + 65y = 715$
$\left(y = -\frac{11}{13}x + 11\right)$. Because the inequality
contains $\le$, use a solid line.
Test Point: $(0,0)$: $0 + 0 \le 715$ is true. Shade the
side containing $(0,0)$.

Graph the inequality $y \le 2$.
Boundary line: $y = 2$. Because the inequality
contains $\le$, use a solid line.
Test Point: $(0,0)$: $0 \le 2$ is true. Shade the side
containing $(0,0)$.

The overlapping shaded region (that is, the
shaded region in the graph below) is the solution
to the system of linear inequalities.

The corner points are $(0,0)$, $(13,0)$, $\left(\frac{117}{11},2\right)$
and $(0,2)$.

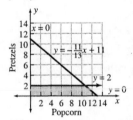

Chapter 7 Test

1. 3x - y = 400,
   3x = 400 + y
   y = 3x − 400

2. x - 2y = 3,
   x = 3 + 2y,
   2y = x - 3,
   $y = \dfrac{x}{2} - \dfrac{3}{2}$

3. a + 2b = 6, a = 6 - 2b
   3a - b = -17,
   3(6 - 2b) - b = -17,
   18 - 6b - b = -17
   -7b = -35,
   b = 5
   a = 6 - 2(5), a = -4
   The solution is $a = -4, b = 5$.

4. Solve both equations for b,
   b = 2 - 5m, b = -1 − m
   2 - 5m = -1 - m

3 = 4m
m = 0.75
b = -1 - 0.75, b = -1.75
The solution is $b = -1.75, m = 0.75$.

5. Solve both equations for y,
   y = 5 - x, y = -x + 5
   5 - x = -x + 5, always true, infinite number of
   solutions.

6. Solve both equations for y,
   y = 2x - 1, y = 3 + 2x
   2x - 1 = 3 + 2x,
   0 = 4, always false, no solutions

7. Multiply second equation by -2,
   (-2)(5x - 2y) = (47)(-2)
     -10x + 4y = -94
   +   3x - 4y = 3
   ────────────────
            -7x = -91,
             x = 13

   5(13) - 2y = 47, -2y = -18, y = 9
   The solution is $x = 13, y = 9$.

8. Solve second equation for x,
   x = 900 + y,
   3(900 + y) + 2y = 5700, 2700 + 5y = 5700,
   5y = 3000, y = 600
   x = 900 + 600, x = 1500
   The solution is $x = 1500, y = 600$.

9. The graphs are coincident lines.

10. The lines are parallel.

11. (6, 2) is the point of intersection of
    y = 8 - x and y = 4 - $\frac{1}{3}$ x.

    Substituting the intersection coordinates into the
    equations will make both equations true.

12.
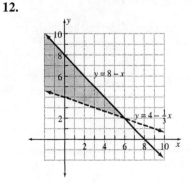

13. $1 < 2(7) + 2$, true
$1 \leq 8 - 7$, true
$(7, 1)$ is in the solution set because it makes both inequalities true.

14. Let c = legs on a caterpillar and b = legs on a butterfly
$c = b + 10$, $6c + 8b = 144$
$6(b + 10) + 8b = 144$,
$6b + 60 + 8b = 144$
$14b = 84$,
$b = 6$,
$c = 6 + 10$, $c = 16$
The caterpillar has 16 legs and the butterfly has 6 legs.

15. Let p = calories per peanut and c = calories per cashew
$16p + 5c = 135$, $20p + 25c = 405$
$(-5)(16p + 5c) = (135)(-5)$
$\quad -80p - 25c = -675$
$\underline{+\ 20p + 25c = 405}$
$\qquad\quad -60p = -270$
$\qquad\qquad\quad p = 4.5$
$20(4.5) + 25c = 405$,
$25c = 315$,
$c = 12.6$
Peanuts have 4.5 calories and cashews have 12.6 calories.

16. Let r = the speed of the dolphin and c = the speed of the current
$3(r + c) = 135$, $4(r - c) = 100$
$\quad r + c = 45$
$\underline{+\ r - c = 25}$
$\quad 2r = 70$,
$\qquad r = 35$

$35 + c = 45$
$c = 10$
The speed of the dolphin is 35 mph and the speed of the current is 10 mph.

17. Let x be the measure of the smaller angle and y be the measure of the larger angle.
$x + y = 90$
$y - 12 = 3x + 24$
Solve the second equation for y.
$y = 3x + 36$
Substitute this result into the first equation and solve for x.

$x + y = 90$
$x + 3x + 36 = 90$
$4x + 36 = 90$
$4x = 54$
$x = 13.5$
Substitute $x = 13.5$ into the equation for y above.
$y = 3x + 36$
$y = 3(13.5) + 36$
$y = 76.5$
The larger angle measures $76.5°$ and the smaller angle measures $13.5°$.

18. $x + y \leq 216$, $y \leq 50$, $x \geq 0$, $y \geq 0$

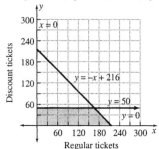

19. If a variable has a coefficient of 1 or -1 use substitution.

20. $y - z = 3$
$x + y = 2$
$x - z = -5$
Solve the second equation for y.
$y = -x + 2$
Solve the third equation for z.
$z = x + 5$
Substitute these results into the first equation and solve for x.
$$y - z = 3$$
$$(-x + 2) - (x + 5) = 3$$
$$-x + 2 - x - 5 = 3$$
$$-2x - 3 = 3$$
$$-2x = 6$$
$$x = -3$$
Substitute $x = -3$ into the equations for y and z above.
$y = -x + 2 = -(-3) + 2 = 5$
$z = x + 5 = -3 + 5 = 2$
The solution is $(-3, 5, 2)$.

Chapter 8

Exercises 8.1

1. $x^2 = 3^2 + 5^2$

    $x^2 = 9 + 25$

    $x^2 = 34$

    $x = \sqrt{34} \approx 5.83$

3. $x^2 = 6^2 + 9^2$

    $x^2 = 36 + 81$

    $x^2 = 117$

    $x = \sqrt{117} \approx 10.82$

5. $5^2 + x^2 = 8^2$

    $25 + x^2 = 64$

    $x^2 = 39$

    $x = \sqrt{39} \approx 6.24$

7. $7^2 = 49$, $8^2 = 64$, $9^2 = 81$

    $49 + 64 \neq 81$

    Since the sum of the squares of the two shorter sides do not equal the square of the longest side, the sides do not form a right triangle.

9. $12^2 = 144$, $16^2 = 256$, $20^2 = 400$

    $144 + 256 = 400$

    The sum of the squares of the two smaller sides equals the square of the longest side. The sides form a right triangle.

11. $7^2 = 49$, $24^2 = 576$, $25^2 = 625$

    $49 + 576 = 625$

    The sum of the squares of the two smaller sides equals the square of the longest side. The sides form a right triangle.

13. **a.** $2^2 = 4$, $2.1^2 = 4.41$, $2.9^2 = 8.41$

    Since $2^2 + 2.1^2 = 2.9^2$, the set forms a Pythagorean triple.

    b. $1^2 = 1$, $\left(\sqrt{3}\right)^2 = 3$, $2^2 = 4$

    Since $1^2 + \left(\sqrt{3}\right)^2 = 2^2$, the set forms a Pythagorean triple.

c. $1^2 = 1$, $2^2 = 4$, $3^2 = 9$

The numbers cannot be arranged so that $a^2 + b^2 = c^2$. The set is not a Pythagorean triple.

d. $\left(\dfrac{3}{7}\right)^2 = \dfrac{9}{49}$, $\left(\dfrac{4}{7}\right)^2 = \dfrac{16}{49}$, $\left(\dfrac{5}{7}\right)^2 = \dfrac{25}{49}$

Since $\left(\dfrac{3}{7}\right)^2 + \left(\dfrac{4}{7}\right)^2 = \left(\dfrac{5}{7}\right)^2$, the set forms a Pythagorean triple.

15. **a.** $\left(\sqrt{7}\right)^2 = 7$, $3^2 = 9$, $4^2 = 16$

    Since $\left(\sqrt{7}\right)^2 + 3^2 = 4^2$, the set forms a Pythagorean triple.

    b. $(-5)^2 = 25$, $(-12)^2 = 144$, $(-13)^2 = 169$

    Since $(-5)^2 + (-12)^2 = (-13)^2$, the set forms a Pythagorean triple. It does not satisfy the Pythagorean theorem because the sides of a triangle cannot have negative length.

    c. $\left(\sqrt{13}\right)^2 = 13$, $6^2 = 36$, $7^2 = 49$

    Since $\left(\sqrt{13}\right)^2 + 6^2 = 7^2$, the set forms a Pythagorean triple.

    d. $\left(\dfrac{3}{4}\right)^2 = \dfrac{9}{16}$, $\left(\dfrac{4}{5}\right)^2 = \dfrac{16}{25}$, $\left(\dfrac{5}{6}\right)^2 = \dfrac{25}{36}$

    The numbers cannot be arranged so that $a^2 + b^2 = c^2$. The set is not a Pythagorean triple.

17.

Leg	Leg	Hypotenuse
3	4	5
$6 = 2 \cdot 3$	$2 \cdot 4 = 8$	$2 \cdot 5 = 10$
$6 \cdot 3 = 18$	$6 \cdot 4 = 24$	$30 = 6 \cdot 5$
$3 \cdot 3 = 9$	$12 = 4 \cdot 3$	$5 \cdot 3 = 15$
$1 = 3 \cdot \dfrac{1}{3}$	$4 \cdot \dfrac{1}{3} = \dfrac{4}{3}$	$5 \cdot \dfrac{1}{3} = \dfrac{5}{3}$

19. $\dfrac{x}{6} = \dfrac{28}{8}$ $\dfrac{w}{10} = \dfrac{28}{8}$

$8x = 6(28)$ $8w = 10(28)$

$8x = 168$ $8w = 280$

$x = 21$ $w = 35$

Since $6^2 + 8^2 = 10^2$ and $21^2 + 28^2 = 35^2$, the two triangles are right triangles.

21. $x^2 + x^2 = 8^2$

$\quad 2x^2 = 64$

$\quad\quad x^2 = 32$

$\quad\quad\quad x = \sqrt{32} \approx 5.7$

23. $x^2 + (5x)^2 = 52^2$

$\quad x^2 + 25x^2 = 2704$

$\quad\quad 26x^2 = 2704$

$\quad\quad\quad x^2 = 104$

$\quad\quad\quad\quad x = \sqrt{104} \approx 10.2$

25. $x^2 + (2x)^2 = 15^2$

$\quad x^2 + 4x^2 = 225$

$\quad\quad 5x^2 = 225$

$\quad\quad\quad x^2 = 45$

$\quad\quad\quad\quad x = \sqrt{45} \approx 6.7$

27. Let x be the length of the ladder.

$12^2 + 3^2 = x^2$

$144 + 9 = x^2$

$\quad 153 = x^2$

$\quad \sqrt{153} = x$

$\quad\quad x \approx 12.4$

The ladder would be about 12.4 feet long.

29. $\sqrt{9^2 + 2.25^2} = \sqrt{81 + 5.06} \approx 9.3$

The ladder would need to be about 9.3 ft long.

31. $\dfrac{14}{b} = \dfrac{4}{1}, \quad b = \dfrac{14}{4}, \quad b = 3.5$

Safe position is 3.5 ft from the base of the ladder to the wall.

$\sqrt{14^2 + 3.5^2} = \sqrt{196 + 12.25} \approx 14.4$

The ladder would need to be about 14.4 ft long.

33. From the safe ladder ratio, if the base is x the height would be 4x. Now using the Pythagorean theorem:

$12^2 = x^2 + (4x)^2$

$144 = x^2 + 16x^2$

$17x^2 = 144$

$\quad x^2 = \dfrac{144}{17}$

$\quad\quad x = \sqrt{\dfrac{144}{17}}$

$\quad\quad x \approx 2.91$ feet

$4x \approx 4(2.91)$

$4x \approx 11.64$ ft

$(0.91)(12) \approx 11$

$(0.64)(12) \approx 8$

base is $\approx$ 2.91 ft or 2 ft 11 in.

height is $\approx$ 11.64 ft or 11 ft 8 in.

35. $18^2 = x^2 + (4x)^2$

$324 = x^2 + 16x^2$

$17x^2 = 324$

$\quad x^2 = \dfrac{324}{17}$

$\quad\quad x = \sqrt{\dfrac{324}{17}} \approx 4.37$ ft

$4x \approx 4(4.37)$

$4x \approx 17.48$ ft

$(0.37)(12) \approx 4$

$(0.48)(12) \approx 6$

base is $\approx$ 4.37 ft or 4 ft 4 in.

height is $\approx$ 17.48 ft or 17 ft 6 in.

37. $c^2 = 250^2 + 30^2$

$\quad c = \sqrt{250^2 + 30^2}$

$\quad\quad = \sqrt{63400}$

$\quad\quad \approx 251.8$

The plane will have traveled about 251.8 miles.

250

30

39. $c^2 = 200^2 + 32^2$

$\quad c = \sqrt{200^2 + 32^2}$

$\quad\quad = \sqrt{41024}$

$\quad\quad \approx 202.5$

The plane would fly about 202.5 miles.

32

200

41. a. The length of the roof is 20 feet, the same as the length of the house. The width of one side of the roof is given as 13 feet. The area of one side is given by:

$A = l \cdot w = 20 \cdot 13 = 260 \text{ ft}^2$

The total area of the roof is twice this amount: 520 ft^2.

b. The right triangle formed from the peak of the roof has a base of $24 \div 2 = 12$ and a hypotenuse of 13.

$a^2 + b^2 = c^2$

$a^2 + 12^2 = 13^2$

$a^2 + 144 = 169$

$\quad a^2 = 25$

$\quad\quad a = 5$

The roof rises 5 feet over a distance of 12 feet. Thus, the slope is $\dfrac{5}{12}$.

c. The total height, h, is 10 + the height of the triangle. Thus, the height of the house is 15 feet.

43. With a slope of 3 to 14 one edge of the roof line is $\sqrt{3^2 + 14^2} = \sqrt{205} \approx 14.3$ ft.

Area of roof is $(14.3)(40)(2) \approx 1146 \text{ ft}^2$.

45. One edge of roof line is

$\sqrt{9^2 + 14^2} = \sqrt{277} \approx 16.6$ ft.

Area of roof is $(16.6)(40)(2) \approx 1332 \text{ ft}^2$.

47. $x^2 + 4x^2 = 16^2$

$\quad\quad 5x^2 = 256$

$\quad\quad\quad x^2 = \dfrac{256}{5}$

$\quad\quad\quad\quad x = \sqrt{\dfrac{256}{5}} \approx 7.16$ ft

$4x = 4(7.16) = 28.64$ ft which is well beyond the length of the ladder.

Exercises 8.2

1. a. -3

  b. ab

  c. $4x^2$

  d. 2

  e. x

  f. $x + 2$

3. a. $\sqrt{81} = 9$

  b. $\sqrt{15} \approx 3.873$, I

  c. $\sqrt{25} = 5$

  d. $\sqrt{2.25} = 1.5$

5. a. $\sqrt{400} = 20$

  b. $\sqrt{35} \approx 5.916$, I

  c. $\sqrt{16} = 4$

  d. $\sqrt{0.01} = 0.1$

7. a. $\sqrt{24} = 2\sqrt{6} \approx 4.899$, I

    b. $\sqrt{144} = 12$

    c. $\sqrt{6.25} = 2.5$

    d. $\sqrt{0.1} \approx 0.316$, I

9. a. Since $8^2 = 64 < 80 < 81 = 9^2$, we can say that $\sqrt{80}$ is between 8 and 9.

    b. Since $7^2 = 49 < 54 < 64 = 8^2$, we can say that $\sqrt{54}$ is between 7 and 8.

    c. Since $14^2 = 196 < 221 < 225 = 15^2$, we can say that $\sqrt{221}$ is between 14 and 15.

    d. Since $4^2 = 16 < 18 < 25 = 5^2$, we can say that $\sqrt{18}$ is between 4 and 5.

11. a. $\sqrt{-36}$ is not a real number.

    b. $-\sqrt{81} = -9$

    c. $\pm\sqrt{144} = \pm 12$

13. a. $\sqrt{49} = 7$

    b. $-\sqrt{225} = -15$

    c. $\pm\sqrt{400} = \pm 20$

15. a. $\sqrt{5} \cdot \sqrt{3} = \sqrt{5 \cdot 3} = \sqrt{15}$

    b. $\left(3\sqrt{5}\right)^2 = 3^2 \cdot \left(\sqrt{5}\right)^2 = 9 \cdot 5 = 45$

    c. $\sqrt{16 \cdot 3} = \sqrt{16} \cdot \sqrt{3} = 4\sqrt{3}$

17. a. $\left(3\sqrt{7}\right)^2 = 3^2 \left(\sqrt{7}\right)^2 = 9 \cdot 7 = 63$

    b. $\sqrt{13} \cdot \sqrt{13} = \sqrt{13 \cdot 13} = \sqrt{13^2} = 13$

    c. $\sqrt{36 \cdot 2} = \sqrt{36} \cdot \sqrt{2} = 6\sqrt{2}$

19. a. $\sqrt{3} \cdot \sqrt{27} = \sqrt{3 \cdot 27} = \sqrt{81} = 9$

    b. $\left(2\sqrt{5}\right)^2 = 2^2 \left(\sqrt{5}\right)^2 = 4 \cdot 5 = 20$

    c. $\sqrt{18} = \sqrt{9 \cdot 2} = \sqrt{9} \cdot \sqrt{2} = 3\sqrt{2}$

21. a. $\sqrt{a} \cdot \sqrt{a} = \left(\sqrt{a}\right)^2 = a$

    b. $\sqrt{b^2} = b$

    c. $\sqrt{121a^4} = \sqrt{121} \cdot \sqrt{a^4} = 11a^2$

23. a. $\sqrt{32a} \cdot \sqrt{2a} = \sqrt{32a \cdot 2a} = \sqrt{64a^2} = 8a$

    b. $\sqrt{2x} \cdot \sqrt{2x} = \left(\sqrt{2x}\right)^2 = 2x$

    c. $\sqrt{16b^2} = \sqrt{16} \cdot \sqrt{b^2} = 4b$

25. a. $\sqrt{\dfrac{x^2}{9}} = \dfrac{\sqrt{x^2}}{\sqrt{9}} = \dfrac{x}{3}$

    b. $\sqrt{\dfrac{4}{25}} = \dfrac{\sqrt{4}}{\sqrt{25}} = \dfrac{2}{5}$

    c. $\sqrt{\dfrac{45}{5}} = \sqrt{9} = 3$

27. a. $\sqrt{\dfrac{28x}{7x^3}} = \sqrt{\dfrac{7 \cdot 4 \cdot x}{7 \cdot x \cdot x^2}} = \sqrt{\dfrac{4}{x^2}} = \dfrac{2}{x}$

    b. $\sqrt{\dfrac{3x^2}{27}} = \sqrt{\dfrac{3 \cdot x^2}{3 \cdot 9}} = \sqrt{\dfrac{x^2}{9}} = \dfrac{x}{3}$

    c. $\sqrt{\dfrac{8a^4}{32}} = \sqrt{\dfrac{8 \cdot a^4}{8 \cdot 4}} = \sqrt{\dfrac{a^4}{4}} = \dfrac{a^2}{2}$

29. a. $\sqrt{\dfrac{8x^2y}{2y}} = \sqrt{\dfrac{2 \cdot 4 \cdot x^2 \cdot y}{2 \cdot y}} = \sqrt{4x^2} = 2x$

    b. $\sqrt{\dfrac{2x^4}{50y^2}} = \sqrt{\dfrac{2 \cdot x^4}{2 \cdot 25 \cdot y^2}} = \sqrt{\dfrac{x^4}{25y^2}} = \dfrac{x^2}{5y}$

    c. $-\sqrt{\dfrac{3xy^4}{27x}} = -\sqrt{\dfrac{3 \cdot x \cdot y^4}{3 \cdot 9 \cdot x}} = -\sqrt{\dfrac{y^4}{9}} = -\dfrac{y^2}{3}$

d. $\dfrac{\sqrt{8xy^5}}{\sqrt{2x^3y}} = \dfrac{\sqrt{4 \cdot 2 \cdot x \cdot y^4 \cdot y}}{\sqrt{2 \cdot x^2 \cdot x \cdot y}}$

$\phantom{\dfrac{\sqrt{8xy^5}}{\sqrt{2x^3y}}} = \dfrac{2y^2\sqrt{2xy}}{x\sqrt{2xy}}$

$\phantom{\dfrac{\sqrt{8xy^5}}{\sqrt{2x^3y}}} = \dfrac{2y^2}{x}$

31. a. $d = \sqrt{(4-2)^2 + (9-3)^2}$

$ = \sqrt{(2)^2 + (6)^2}$

$ = \sqrt{4+36}$

$ = \sqrt{40}$

$ = 2\sqrt{10}$

b. $m = \dfrac{9-3}{4-2} = \dfrac{6}{2} = 3$

c. $y = mx + b$

$3 = 3(2) + b$

$3 = 6 + b$

$-3 = b$

$y = 3x - 3$

33. a. $d = \sqrt{(5-2)^2 + (-1-2)^2}$

$ = \sqrt{(3)^2 + (3)^2}$

$ = \sqrt{9+9}$

$ = \sqrt{18}$

$ = 3\sqrt{2}$

b. $m = \dfrac{-1-2}{5-2} = \dfrac{-3}{3} = -1$

c. $y = mx + b$

$2 = (-1)(2) + b$

$2 = -2 + b$

$4 = b$

$y = -x + 4$

35. a. $d = \sqrt{(4-(-3))^2 + (2-3)^2}$

$ = \sqrt{(7)^2 + (-1)^2}$

$ = \sqrt{49+1}$

$ = \sqrt{50}$

$ = 5\sqrt{2}$

b. $m = \dfrac{2-3}{4-(-3)} = \dfrac{-1}{7} = -\dfrac{1}{7}$

c. $y = mx + b$

$3 = -\dfrac{1}{7}(-3) + b$

$3 = \dfrac{3}{7} + b$

$\dfrac{18}{7} = b$

$y = -\dfrac{1}{7}x + \dfrac{18}{7}$

37. a. $d = \sqrt{(3-(-3))^2 + (-3-(-1))^2}$

$ = \sqrt{(6)^2 + (-2)^2}$

$ = \sqrt{36+4}$

$ = \sqrt{40}$

$ = 2\sqrt{10}$

b. $m = \dfrac{-3-(-1)}{3-(-3)} = \dfrac{-2}{6} = -\dfrac{1}{3}$

c. $y = mx + b$

$-1 = \left(-\dfrac{1}{3}\right)(-3) + b$

$-1 = 1 + b$

$-2 = b$

$y = -\dfrac{1}{3}x - 2$

39. Let $A = (3,4)$, $B = (0,1)$, and $C = (6,1)$.

$AB = \sqrt{(0-3)^2 + (1-4)^2}$

$ = \sqrt{9+9}$

$ = \sqrt{18}$

$$BC = \sqrt{(6-0)^2 + (1-1)^2}$$
$$= \sqrt{36+0}$$
$$= 6$$
$$AC = \sqrt{(6-3)^2 + (1-4)^2}$$
$$= \sqrt{9+9}$$
$$= \sqrt{18}$$
$$\left(\sqrt{18}\right)^2 + \left(\sqrt{18}\right)^2 = (6)^2$$
$$18+18 = 36$$
$$36 = 36 \text{ true}$$

Since $AB = AC$ and the sides form a Pythagorean triple, this is an isosceles right triangle.

41. Let $A = (6,5)$, $B = (4,2)$, and $C = (8,2)$.

$$AB = \sqrt{(4-6)^2 + (2-5)^2}$$
$$= \sqrt{4+9}$$
$$= \sqrt{13}$$
$$BC = \sqrt{(8-4)^2 + (2-2)^2}$$
$$= \sqrt{16+0}$$
$$= 4$$
$$AC = \sqrt{(8-6)^2 + (2-5)^2}$$
$$= \sqrt{4+9}$$
$$= \sqrt{13}$$
$$\left(\sqrt{13}\right)^2 + \left(\sqrt{13}\right)^2 = 4^2$$
$$13+13 = 16$$
$$26 = 16 \text{ false}$$

$AB = AC$, but the sides do not form a Pythagorean triple. This is an isosceles triangle.

43. Let $A = (4,8)$, $B = (1,6)$, and $C = (5,0)$.

$$AB = \sqrt{(1-4)^2 + (6-8)^2}$$
$$= \sqrt{9+4}$$
$$= \sqrt{13}$$
$$BC = \sqrt{(5-1)^2 + (0-6)^2}$$
$$= \sqrt{16+36}$$
$$= \sqrt{52}$$

$$AC = \sqrt{(5-4)^2 + (0-8)^2}$$
$$= \sqrt{1+64}$$
$$= \sqrt{65}$$
$$\left(\sqrt{13}\right)^2 + \left(\sqrt{52}\right)^2 = \left(\sqrt{65}\right)^2$$
$$13+52 = 65$$
$$65 = 65 \text{ true}$$

None of the sides are equal, but they form a Pythagorean triple. This is a right triangle.

45. False. If $0 < x < 1$, we would have $\sqrt{x} > x$. For example, $\sqrt{\dfrac{1}{4}} = \dfrac{\sqrt{1}}{\sqrt{4}} = \dfrac{1}{2}$.

47. $\sqrt{\dfrac{a}{b}} = \left(\dfrac{a}{b}\right)^{1/2} = \dfrac{a^{1/2}}{b^{1/2}} = \dfrac{\sqrt{a}}{\sqrt{b}}$

49.
  a. $25^0 = 1$

  b. $25^{-1} = \dfrac{1}{25}$

  c. $25^{1/2} = \sqrt{25} = 5$

  d. $25^{0.5} = 25^{1/2} = \sqrt{25} = 5$

51.
  a. $9^{1/2} = \sqrt{9} = 3$

  b. $9^0 = 1$

  c. $9^{0.5} = 9^{1/2} = \sqrt{9} = 3$

  d. $9^{-1} = \dfrac{1}{9}$

53.
  a. $\left(\dfrac{1}{4}\right)^{-1} = \left(\dfrac{4}{1}\right)^1 = 4$

  b. $\left(\dfrac{1}{4}\right)^0 = 1$

  c. $\left(\dfrac{1}{4}\right)^{1/2} = \sqrt{\dfrac{1}{4}} = \dfrac{\sqrt{1}}{\sqrt{4}} = \dfrac{1}{2}$

  d. $\left(\dfrac{1}{4}\right)^{0.5} = \left(\dfrac{1}{4}\right)^{1/2} = \sqrt{\dfrac{1}{4}} = \dfrac{\sqrt{1}}{\sqrt{4}} = \dfrac{1}{2}$

55. a. $(0.25)^{-1} = \left(\dfrac{1}{4}\right)^{-1} = \left(\dfrac{4}{1}\right)^{1} = 4$

b. $(0.01)^{0.5} = \left(\dfrac{1}{100}\right)^{1/2} = \sqrt{\dfrac{1}{100}} = \dfrac{1}{10} = 0.1$

c. $(6.25)^{0.5} = \left(\dfrac{625}{100}\right)^{1/2} = \sqrt{\dfrac{625}{100}} = \dfrac{5}{2} = 2.5$

d. $(0.25)^{0.5} = \left(\dfrac{1}{4}\right)^{1/2} = \sqrt{\dfrac{1}{4}} = \dfrac{1}{2} = 0.5$

e. $(0.02)^{-1} = \left(\dfrac{2}{100}\right)^{-1} = \left(\dfrac{100}{2}\right)^{1} = 50$

f. $(0.05)^{-1} = \left(\dfrac{5}{100}\right)^{-1} = \left(\dfrac{100}{5}\right)^{1} = 20$

Exercises 8.3

1. a. $\sqrt{ab^2} = \sqrt{a}\sqrt{b^2} = |b|\sqrt{a}$, $a \geq 0$

b. $\sqrt{a^2 b} = \sqrt{a^2}\sqrt{b} = |a|\sqrt{b}$, $b \geq 0$

c. $\sqrt{a^2 b^2} = \sqrt{(ab)^2} = |ab|$

3. a. $\sqrt{49x^2} = \sqrt{49}\sqrt{x^2} = 7|x|$

b. $\sqrt{121y^2} = \sqrt{121}\sqrt{y^2} = 11|y|$

c. $\sqrt{x^6} = \sqrt{\left(x^3\right)^2} = |x^3|$

5. a. $\sqrt{p^1 q^2} = \sqrt{p}\sqrt{q^2} = |q|\sqrt{p}$, $p \geq 0$

b. $\sqrt{p^4} = p^2$

c. $\sqrt{r^6 \cdot s^8} = \sqrt{r^6} \cdot \sqrt{s^8}$

$\qquad = \sqrt{\left(r^3\right)^2} \cdot \sqrt{\left(s^4\right)^2}$

$\qquad = |r^3| s^4$

7. a. $\sqrt{a \cdot b^4} = \sqrt{a} \cdot \sqrt{b^4} = b^2\sqrt{a}$, $a \geq 0$

b. $\sqrt{b \cdot c^2} = \sqrt{b} \cdot \sqrt{c^2} = |c|\sqrt{b}$, $b \geq 0$

9. $\sqrt{x^2 y} = |x|\sqrt{y}$ needs an absolute value while $\sqrt{x^4 y} = x^2\sqrt{y}$ does not because $x^2 \geq 0$.

11. $f(x) = \sqrt{4-x}$

$f(-4) = \sqrt{4-(-4)} = \sqrt{8} = 2\sqrt{2}$

$f(-1) = \sqrt{4-(-1)} = \sqrt{5}$

$f(0) = \sqrt{4-0} = \sqrt{4} = 2$

$f(4) = \sqrt{4-4} = \sqrt{0} = 0$

$f(6) = \sqrt{4-6} = \sqrt{-2}$ not a real number.

13. $f(x) = \sqrt{3-x}$

$f(-4) = \sqrt{3-(-4)} = \sqrt{7}$

$f(-1) = \sqrt{3-(-1)} = \sqrt{4} = 2$

$f(0) = \sqrt{3-0} = \sqrt{3}$

$f(4) = \sqrt{3-4} = \sqrt{-1}$ not a real number

$f(6) = \sqrt{3-6} = \sqrt{-3}$ not a real number

15. $y = \sqrt{x+3}$

When $x = -2$, we have $y = \sqrt{-2+3} = \sqrt{1} = 1$. Therefore, $y = \sqrt{x+3}$ is defined for $x = -2$.

17. $y = \sqrt{-2x}$ is defined for all values of x that make the radicand nonnegative. That is, we need $-2x \geq 0$ or $x \leq 0$. So, $y = \sqrt{-2x}$ is defined for all values of $x \leq 0$.

19.

x	$y = \sqrt{x+4}$
-4	$\sqrt{-4+4} = 0$
-2	$\sqrt{-2+4} = \sqrt{2}$
0	$\sqrt{0+4} = 2$
2	$\sqrt{2+4} = \sqrt{6}$
4	$\sqrt{4+4} = 2\sqrt{2}$

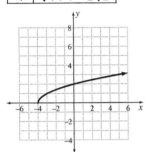

21.

x	$y = \sqrt{x-2}$
-4	$\sqrt{-4-2} = \sqrt{-6}$ not a real number
-2	$\sqrt{-2-2} = \sqrt{-4}$ not a real number
0	$\sqrt{0-2} = \sqrt{-2}$ not a real number
2	$\sqrt{2-2} = 0$
4	$\sqrt{4-2} = \sqrt{2}$

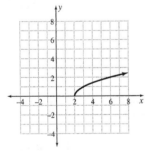

23.

x	$y = \sqrt{2x}$
-4	$\sqrt{2(-4)} = \sqrt{-8}$ not a real number
-2	$\sqrt{2(-2)} = \sqrt{-4}$ not a real number
0	$\sqrt{2(0)} = 0$
2	$\sqrt{2(2)} = 2$
4	$\sqrt{2(4)} = 2\sqrt{2}$

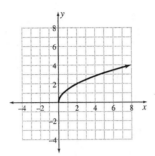

25. The expression is defined when the radicand is nonnegative.

   a. $x - 1 \geq 0$

   $x \geq 1$

   b. $x + 3 \geq 0$

   $x \geq -3$

27. The expression is defined when the radicand is nonnegative.

   a. $4 - x \geq 0$

   $-x \geq -4$

   $x \leq 4$

   b. $3 - x \geq 0$

   $-x \geq -3$

   $x \leq 3$

29. The solution to the equation is the value for x where the graphs cross. The graphs cross at $x = 6$, so the solution to the system is $x = 6$.

31. $y = \sqrt{x+3}$ is defined when $x \geq -3$.

   $$\sqrt{x+3} = 1$$

   $$\left(\sqrt{x+3}\right)^2 = 1^2$$

   $$x + 3 = 1$$

   $$x = -2$$

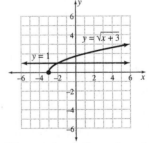

The graph confirms the solution. The two graphs cross when $x = -2$.

33. $y = \sqrt{3-x}$ is defined when $3 - x \geq 0$, or $x \leq 3$.

$$\sqrt{3-x} = 2$$
$$\left(\sqrt{3-x}\right)^2 = 2^2$$
$$3 - x = 4$$
$$-x = 1$$
$$x = -1$$

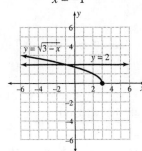

The graph confirms the solution. The two graphs cross when $x = -1$.

35.
$$\sqrt{x-2} = 3$$
$$\left(\sqrt{x-2}\right)^2 = 3^2$$
$$x - 2 = 9$$
$$x = 11$$

The radical expression is defined when $x - 2 \geq 0$, or $x \geq 2$.

37.
$$\sqrt{2x+2} = 4$$
$$\left(\sqrt{2x+2}\right)^2 = 4^2$$
$$2x + 2 = 16$$
$$2x = 14$$
$$x = 7$$

The radical expression is defined when $2x + 2 \geq 0$. That is,
$$2x + 2 \geq 0$$
$$2x \geq -2$$
$$x \geq -1$$

39.
$$\sqrt{3x-3} = 6$$
$$\left(\sqrt{3x-3}\right)^2 = 6^2$$
$$3x - 3 = 36$$
$$3x = 39$$
$$x = 13$$

The radical expression is defined when $3x - 3 \geq 0$. That is,
$$3x - 3 \geq 0$$
$$3x \geq 3$$
$$x \geq 1$$

41.
$$\sqrt{5-x} = 3$$
$$\left(\sqrt{5-x}\right)^2 = 3^2$$
$$5 - x = 9$$
$$-x = 4$$
$$x = -4$$

The radical expression is defined when $5 - x \geq 0$, or $x \leq 5$.

43. $\sqrt{x-2} = -1$

This equation has no real solution since a (principal) square root always yields a nonnegative result.
The radical expression is defined when $x - 2 \geq 0$, or $x \geq 2$.

45. The graph of $y = \sqrt{x-2}$ will not intersect the graph of $y = -1$.

47. $d \approx \sqrt{\dfrac{3h}{2}} = \sqrt{\dfrac{3(1454)}{2}} = \sqrt{2181} \approx 46.7$

From the top of the Sears Tower, you could see roughly 46.7 miles.

49. $d \approx \sqrt{\dfrac{3h}{2}} = \sqrt{\dfrac{3(1002)}{2}} = \sqrt{1503} \approx 38.8$

From the top of the Texas Commerce Tower, you could see roughly 38.8 miles.

51. $d \approx \sqrt{\dfrac{3h}{8}} = \sqrt{\dfrac{3(24)}{8}} = \sqrt{9} = 3$

From a height of 24 feet on the moon, you can see roughly 3 miles.

53. $d \approx \sqrt{\dfrac{3h}{8}} = \sqrt{\dfrac{3(96)}{8}} = \sqrt{36} = 6$

From a height of 96 feet on the moon, you can see roughly 6 miles.

55. a.

$$\sqrt{\frac{3h}{2}} = 30$$

$$\left(\sqrt{\frac{3h}{2}}\right)^2 = 30^2$$

$$\frac{3h}{2} = 900$$

$$3h = 1800$$

$$h = 600$$

To see 600 miles on Earth, we would need to be at a height of 600 feet.

b.

$$\sqrt{\frac{3h}{8}} = 30$$

$$\left(\sqrt{\frac{3h}{8}}\right)^2 = 30^2$$

$$\frac{3h}{8} = 900$$

$$3h = 7200$$

$$h = 2400$$

To see 600 miles on the moon, we would need to be at a height of 2400 feet.

c. We would need to be 4 times as high on the moon as on Earth.

d. The moon's radius is smaller and it has a greater curvature.

57. $t = \sqrt{\dfrac{2h}{g}} = \sqrt{\dfrac{2(1821)}{32.2}} = \sqrt{\dfrac{3642}{32.2}} \approx 10.6$

An object dropped from the top of the C.N. Tower would take about 10.6 seconds to reach the ground.

$v = gt \approx (32.2)(10.6) \approx 341$

The object would hit the ground traveling at a speed of roughly 341 feet per second.

59. $t = \sqrt{\dfrac{2h}{g}} = \sqrt{\dfrac{2(1063)}{32.2}} = \sqrt{\dfrac{2126}{32.2}} \approx 8.1$

An object dropped from the top of the C&C Plaza would take about 8.1 seconds to hit the ground.

$v = gt \approx (32.2)(8.1) \approx 261$

The object would hit the ground traveling at a speed of roughly 261 feet per second.

61.

$$\sqrt{\frac{2h}{g}} = 12$$

$$\left(\sqrt{\frac{2h}{32.2}}\right)^2 = 12^2$$

$$\frac{2h}{32.2} = 144$$

$$2h = 4636.8$$

$$h = 2318.4$$

It would take an object 12 seconds to fall from the top of a building that is 2318.4 feet tall.

63. 1 mile $= 5280$ feet

$$t = \sqrt{\frac{2h}{g}} = \sqrt{\frac{2(5280)}{32.2}} = \sqrt{\frac{10560}{32.2}} \approx 18.1$$

It would take about 18.1 seconds for an object to fall 1 mile.

65.

$$E = \sqrt{W \cdot R}$$

$$E^2 = \left(\sqrt{W \cdot R}\right)^2$$

$$E^2 = W \cdot R$$

$$\frac{E^2}{W} = R \ \text{ or } \ R = \frac{E^2}{W}$$

67.

$$I = \sqrt{\frac{W}{R}}$$

$$I^2 = \left(\sqrt{\frac{W}{R}}\right)^2$$

$$I^2 = \frac{W}{R}$$

$$R \cdot I^2 = W$$

$$R = \frac{W}{I^2}$$

69.

$$d = \sqrt{\frac{3h}{8}}$$

$$d^2 = \left(\sqrt{\frac{3h}{8}}\right)^2$$

$$d^2 = \frac{3h}{8}$$

$$8d^2 = 3h$$

$$\frac{8d^2}{3} = h \ \text{ or } \ h = \frac{8d^2}{3}$$

71.
$$V_0 = \sqrt{\frac{GM}{R}}$$
$$V_0{}^2 = \left(\sqrt{\frac{GM}{R}}\right)^2$$
$$V_0{}^2 = \frac{GM}{R}$$
$$R \cdot V_0{}^2 = GM$$
$$R = \frac{GM}{V_0{}^2}$$

73.
$$\sqrt{\frac{3h_e}{2}} = \sqrt{\frac{3h_m}{8}}$$
$$\left(\sqrt{\frac{3h_e}{2}}\right)^2 = \left(\sqrt{\frac{3h_m}{8}}\right)^2$$
$$\frac{3h_e}{2} = \frac{3h_m}{8}$$
$$\frac{2}{3}\cdot\frac{3h_e}{2} = \frac{2}{3}\cdot\frac{3h_m}{8}$$
$$h_e = \frac{h_m}{4}$$
$$\frac{h_e}{h_m} = \frac{1}{4}$$

75. We need the absolute value if the resulting expression can be negative.

Mid-Chapter 8 Test

1. a. $4^2 = 16$, $6^2 = 36$, $8^2 = 64$
Since $4^2 + 6^2 \neq 8^2$, the numbers do not represent the sides of a right triangle.

b. $10^2 = 100$, $15^2 = 225$, $20^2 = 400$
Since $10^2 + 15^2 \neq 20^2$, the numbers do not represent the sides of a right triangle.

c. $15^2 = 225$, $20^2 = 400$, $25^2 = 625$
Since $15^2 + 20^2 = 25^2$, the numbers represent the sides of a right triangle.

2. $120^2 + 50^2 = 16900$
$\sqrt{16900} = 130$

In 1 hour, the plane would travel
$(130\text{ mph})(1\text{ h}) = 130$ miles.

3. $c = \sqrt{6^2 + 15^2} = \sqrt{36 + 225} = \sqrt{261}$

4. a. $\sqrt{2}\cdot\sqrt{18} = \sqrt{2\cdot 18} = \sqrt{36} = 6$

b. $\left(3\sqrt{2}\right)^2 = 3^2\left(\sqrt{2}\right)^2 = 9\cdot 2 = 18$

c. $\sqrt{72} = \sqrt{36\cdot 2} = \sqrt{36}\cdot\sqrt{2} = 6\sqrt{2}$

d. $\left(2\sqrt{3}\right)^2 = 2^2\left(\sqrt{3}\right)^2 = 4\cdot 3 = 12$

e. $\sqrt{48} = \sqrt{16\cdot 3} = \sqrt{16}\cdot\sqrt{3} = 4\sqrt{3}$

f. $-\sqrt{4} = -2$

g. $\sqrt{-16}$ is not a real number.

h. $\pm\sqrt{\frac{25}{16}} = \pm\frac{\sqrt{25}}{\sqrt{16}} = \pm\frac{5}{4}$

i. $\frac{\sqrt{2}}{\sqrt{32}} = \sqrt{\frac{2}{32}} = \sqrt{\frac{1}{16}} = \frac{\sqrt{1}}{\sqrt{16}} = \frac{1}{4}$

5. a. $\sqrt{3x}\sqrt{27x^3} = \sqrt{3x\cdot 27x^3}$
$$= \sqrt{81x^4}$$
$$= 9x^2$$

b. $\sqrt{x^2 y^7} = \sqrt{x^2 y^6 \cdot y}$
$$= \sqrt{x^2 y^6}\sqrt{y}$$
$$= xy^3\sqrt{y}$$

c. $\frac{\sqrt{x^3 y^3}}{\sqrt{x^2 y^5}} = \sqrt{\frac{x^3 y^3}{x^2 y^5}} = \sqrt{\frac{x}{y^2}} = \frac{\sqrt{x}}{\sqrt{y^2}} = \frac{\sqrt{x}}{y}$

d. $\sqrt{\dfrac{x^2 y}{xy^5}} = \sqrt{\dfrac{x}{y^4}} = \dfrac{\sqrt{x}}{\sqrt{y^4}} = \dfrac{\sqrt{x}}{y^2}$

6. a. $\sqrt{x^3 y^7} = \sqrt{x^2 y^6 \cdot x \cdot y}$

           $= \sqrt{x^2 y^6} \cdot \sqrt{xy}$

           $= \left| xy^3 \right| \sqrt{xy}$

   b. $\sqrt{\dfrac{3x^2}{12y^4}} = \sqrt{\dfrac{x^2}{4y^4}} = \dfrac{|x|}{2y^2}$

   c. $\sqrt{\dfrac{196x}{x^3}} = \sqrt{\dfrac{196}{x^2}} = \dfrac{\sqrt{196}}{\sqrt{x^2}} = \dfrac{14}{|x|}$

   d. $\sqrt{x^2 + y^2}$ is already simplified.

7. $d = \sqrt{(x_2 - x_1)^2 + (y_2 - y_1)^2}$

      $= \sqrt{(-4 - 2)^2 + (6 - (-3))^2}$

      $= \sqrt{(-6)^2 + (9)^2}$

      $= \sqrt{36 + 81}$

      $= \sqrt{117}$

      $= 3\sqrt{13}$

8. $11^2 = 121$ and $12^2 = 144$. Since 135 is between 121 and 144, we know that $\sqrt{135}$ is between 11 and 12.

9. $I = \sqrt{\dfrac{W}{R}}$

     $I^2 = \left(\sqrt{\dfrac{W}{R}} \right)^2$

     $I^2 = \dfrac{W}{R}$

     $R \cdot I^2 = W$

     $R = \dfrac{W}{I^2}$

10. $\sqrt{x+7} = 4$

     $\left(\sqrt{x+7} \right)^2 = 4^2$

       $x + 7 = 16$

           $x = 9$

The graphs of $y = \sqrt{x+7}$ and $y = 4$ intersect when $x = 9$. The solutions agree. The equation is defined when $x + 7 \geq 0$, or $x \geq -7$.

11. The graph of $y = \sqrt{x}$ shows that the square root of a negative number is undefined because there is no graph in the second or third quadrants (where x is negative).

12. The graph of $y = \sqrt{x}$ shows that the principal square root of a number is nonnegative because there is no graph in quadrants three or four (where y is negative).

Exercises 8.4

1. From the graph, the x-intercept points are $(-3, 0)$ and $(2, 0)$. The y-intercept point is $(0, -6)$. The axis of symmetry is halfway between the two x-intercept points.

     $x = \dfrac{-3 + 2}{2} = -0.5$

     When $x = -0.5$,

     $y = (-0.5)^2 + (-0.5) - 6 = -6.25$

     The vertex is $(-0.5, -6.25)$.

3. From the graph, the x-intercept points are $(0,0)$ and $(5,0)$. The y-intercept is $(0,0)$. The axis of symmetry is halfway between the two x-intercept points.

$$x = \frac{0+5}{2} = 2.5$$

When $x = 2.5$, $y = 5(2.5) - (2.5)^2 = 6.25$.

The vertex is $(2.5, 6.25)$.

5.

x	$y = x^2 - 6x + 7$
-4	$(-4)^2 - 6(-4) + 7 = 47$
-2	$(-2)^2 - 6(-2) + 7 = 23$
0	$(0)^2 - 6(0) + 7 = 7$
2	$(2)^2 - 6(2) + 7 = -1$
4	$(4)^2 - 6(4) + 7 = -1$

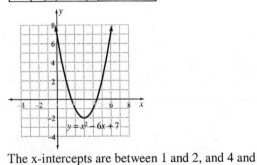

The x-intercepts are between 1 and 2, and 4 and 5. Using a graphing calculator we find the x-intercept points at approximately $(4.4, 0)$ and $(1.6, 0)$. The y-intercept point is $(0, 7)$. The axis of symmetry is $x = 3$. When $x = 3$, $y = (3)^2 - 6(3) + 7 = -2$. The vertex is $(3, -2)$.

7.

x	$y = x^2 + x - 12$
-3	$(-3)^2 + (-3) - 12 = -6$
-1	$(-1)^2 + (-1) - 12 = -12$
0	$(0)^2 + (0) - 12 = -12$
1	$(1)^2 + (1) - 12 = -10$
3	$(3)^2 + (3) - 12 = 0$

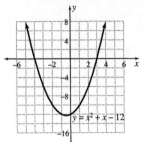

The x-intercept points are $(-4,0)$ and $(3,0)$. The y-intercept point is $(0,-12)$. The axis of symmetry is halfway between the x-intercept points.

$$x = \frac{-4+3}{2} = -0.5$$

When $x = -0.5$,
$$y = (-0.5)^2 + (-0.5) - 12 = -12.25.$$
The vertex is $(-0.5, -12.25)$.

9.

x	$y = x - x^2$
-3	$(-3) - (-3)^2 = -12$
-1	$(-1) - (-1)^2 = -2$
0	$(0) - (0)^2 = 0$
1	$(1) - (1)^2 = 0$
3	$(3) - (3)^2 = -6$

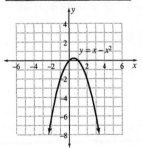

(continued)

The x-intercept points are $(0,0)$ and $(1,0)$. The y-intercept point is $(0,0)$. The axis of symmetry is halfway between the x-intercepts.

$x = \dfrac{0+1}{2} = 0.5$

When $x = 0.5$, $y = (0.5) - (0.5)^2 = 0.25$.

The vertex is $(0.5, 0.25)$.

11.

x	$y = 4x - 2x^2$
-2	$4(-2) - 2(-2)^2 = -16$
-1	$4(-1) - 2(-1)^2 = -6$
0	$4(0) - 2(0)^2 = 0$
1	$4(1) - 2(1)^2 = 2$
2	$4(2) - 2(2)^2 = 0$

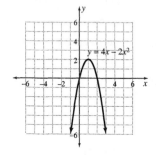

The x-intercept points are $(0,0)$ and $(2,0)$. The y-intercept point is $(0,0)$. The axis of symmetry is halfway between the x-intercepts.

$x = \dfrac{0+2}{2} = 1$

When $x = 1$, $y = 4(1) - 2(1)^2 = 2$. The vertex is $(1,2)$.

13. From the graph, the range is $y \geq -6.25$.

15. From the graph, the range is $y \leq 6.25$.

17. In $y = 2x^2 + 3x + 1$, we have $a = 2$, $b = 3$, and $c = 1$.

19. In $r^2 - 4r + 4 = 0$, $a = 1$, $b = -4$, and $c = 4$.

21. In $y = x^2 - 4$, $a = 1$, $b = 0$, and $c = -4$.

23. In $4t^2 - 8t = 0$, $a = 4$, $b = -8$, and $c = 0$.

25. $\quad 4 = x - x^2$
$x^2 - x + 4 = 0$
$a = 1$, $b = -1$, and $c = 4$.

27. $\quad x^2 = x - 1$
$x^2 - x + 1 = 0$
$a = 1$, $b = -1$, and $c = 1$.

29. In $h = -0.5gt^2 + vt + s$, $a = -0.5g$, $b = v$, and $c = s$.

31. In $A = \pi r^2$, $a = \pi$, $b = 0$, and $c = 0$.

33. $y = 4x^2 + 4x + 1$

35. $y = 9x^2 - 16$

37. $y = 3x^2 + 6x$

39. a. $x^2 + x - 6 = 6$ when $x = -4$ and $x = 3$.

b. $x^2 + x - 6 = -4$ when $x = -2$ and $x = 1$.

c. $x^2 + x - 6 = -8$ has no real solution.

d. $x^2 + x - 6 = 0$ when $x = -3$ and $x = 2$.

41. a. $5x - x^2 = 6$ when $x = 2$ and $x = 3$.

b. $5x - x^2 = 8$ has no real solution.

c. $5x - x^2 = 4$ when $x = 1$ and $x = 4$.

d. $5x - x^2 = 0$ when $x = 0$ and $x = 5$.

43. Solve $h = h_0 - \frac{1}{2}gt^2$ for t.

$h + \frac{1}{2}gt^2 = h_0$

$\frac{1}{2}gt^2 = h_0 - h$

$t^2 = \dfrac{2(h_0 - h)}{g}$

$t = \sqrt{\dfrac{2(h_0 - h)}{g}}$

Half-way is $1612 \div 2 = 806$ ft

$t = \sqrt{\dfrac{2(1612 - 806)}{32.2}}, \quad t \approx 7.1 \text{ sec}$

At the bottom $h = 0$

$t = \sqrt{\dfrac{2(1612 - 0)}{32.2}}, \quad t \approx 10 \text{ sec}$

Exercises 8.5

1. $x^2 = 5$

$\sqrt{x^2} = \sqrt{5}$

$|x| = \sqrt{5}$

$x = \pm\sqrt{5}$

3. $2x^2 = 14$

$x^2 = 7$

$\sqrt{x^2} = \sqrt{7}$

$|x| = \sqrt{7}$

$x = \pm\sqrt{7}$

5. $x^2 = \dfrac{4}{25}$

$\sqrt{x^2} = \sqrt{\dfrac{4}{25}}$

$|x| = \dfrac{2}{5}$

$x = \pm\dfrac{2}{5}$

7. $100x^2 = 4$

$x^2 = \dfrac{4}{100}$

$\sqrt{x^2} = \sqrt{\dfrac{4}{100}}$

$|x| = \dfrac{2}{10}$

$x = \pm\dfrac{1}{5}$

9. $49x^2 - 225 = 0$

$49x^2 = 225$

$x^2 = \dfrac{225}{49}$

$\sqrt{x^2} = \sqrt{\dfrac{225}{49}}$

$|x| = \dfrac{15}{7}$

$x = \pm\dfrac{15}{7}$

11. $36x^2 = 121$

$x^2 = \dfrac{121}{36}$

$\sqrt{x^2} = \sqrt{\dfrac{121}{36}}$

$|x| = \dfrac{11}{6}$

$x = \pm\dfrac{11}{6}$

13. $75x^2 - 27 = 0$

$75x^2 = 27$

$x^2 = \dfrac{27}{75}$

$x^2 = \dfrac{9}{25}$

$\sqrt{x^2} = \sqrt{\dfrac{9}{25}}$

$|x| = \dfrac{3}{5}$

$x = \pm\dfrac{3}{5}$

15. $\dfrac{3}{x} = \dfrac{x}{27}$

$3 \cdot 27 = x \cdot x$

$81 = x^2$

$\sqrt{81} = \sqrt{x^2}$

$9 = |x|$

$\pm 9 = x$

17. $\dfrac{x}{4} = \dfrac{9}{x}$

$x \cdot x = 4 \cdot 9$

$x^2 = 36$

$\sqrt{x^2} = \sqrt{36}$

$|x| = 6$

$x = \pm 6$

19. $(x-4)(x+4) = 0$

$x - 4 = 0 \text{ or } x + 4 = 0$

$x = 4 \text{ or } x = -4$

21. $(2x-1)(3x+2) = 0$

$2x - 1 = 0 \text{ or } 3x + 2 = 0$

$2x = 1 \text{ or } 3x = -2$

$x = \dfrac{1}{2} \text{ or } x = -\dfrac{2}{3}$

23. $(2x-5)(x+2) = 0$

$2x - 5 = 0 \text{ or } x + 2 = 0$

$2x = 5 \text{ or } x = -2$

$x = \dfrac{5}{2} \text{ or } x = -2$

25. $x^2 - 4 = 0$

$(x-2)(x+2) = 0$

$x - 2 = 0 \text{ or } x + 2 = 0$

$x = 2 \text{ or } x = -2$

27. $x^2 + x - 6 = 0$

$(x+3)(x-2) = 0$

$x + 3 = 0 \text{ or } x - 2 = 0$

$x = -3 \text{ or } x = 2$

29. $x^2 - 2x - 15 = 0$

$(x-5)(x+3) = 0$

$x - 5 = 0 \text{ or } x + 3 = 0$

$x = 5 \text{ or } x = -3$

31. $x^2 + 3x = 0$

$x(x+3) = 0$

$x = 0 \text{ or } x + 3 = 0$

$x = 0 \text{ or } x = -3$

33. $2x^2 = -x$

$2x^2 + x = 0$

$x(2x+1) = 0$

$x = 0 \text{ or } 2x + 1 = 0$

$x = 0 \text{ or } 2x = -1$

$x = 0 \text{ or } x = -\dfrac{1}{2}$

35. $x^2 - 4x = 12$

$x^2 - 4x - 12 = 0$

$(x-6)(x+2) = 0$

$x - 6 = 0 \text{ or } x + 2 = 0$

$x = 6 \text{ or } x = -2$

37. $2x^2 = x + 3$

$2x^2 - x - 3 = 0$

$(2x-3)(x+1) = 0$

$2x - 3 = 0 \text{ or } x + 1 = 0$

$2x = 3 \text{ or } x = -1$

$x = \dfrac{3}{2} \text{ or } x = -1$

39. $x^2 = x + 12$

$x^2 - x - 12 = 0$

$(x-4)(x+3) = 0$

$x - 4 = 0 \text{ or } x + 3 = 0$

$x = 4 \text{ or } x = -3$

41. $x^2 + 6 = 7x$

$x^2 - 7x + 6 = 0$

$(x-6)(x-1) = 0$

$x - 6 = 0 \text{ or } x - 1 = 0$

$x = 6 \text{ or } x = 1$

43. $2x^2 + 3x = 5$

$2x^2 - 3x - 5 = 0$

$(2x-5)(x+1) = 0$

$2x - 5 = 0 \text{ or } x + 1 = 0$

$2x = 5 \text{ or } x = -1$

$x = \dfrac{5}{2} \text{ or } x = -1$

45.
$$4x^2 - 25 = 0$$
$$(2x-5)(2x+5) = 0$$
$$2x-5=0 \text{ or } 2x+5=0$$
$$2x=5 \text{ or } 2x=-5$$
$$x=\frac{5}{2} \text{ or } x=-\frac{5}{2}$$

47.
$$3x^2 - 12 = 0$$
$$x^2 - 4 = 0$$
$$(x-2)(x+2) = 0$$
$$x-2=0 \text{ or } x+2=0$$
$$x=2 \text{ or } x=-2$$

49.
$$\frac{x+2}{2} = \frac{5}{x-1}$$
$$(x+2)(x-1) = 2\cdot 5$$
$$x^2 + 2x - x - 2 = 10$$
$$x^2 + x - 12 = 0$$
$$(x+4)(x-3) = 0$$
$$x+4=0 \text{ or } x-3=0$$
$$x=-4 \text{ or } x=3$$

51.
$$\frac{1}{x-3} = \frac{x-2}{2}$$
$$1\cdot 2 = (x-2)(x-3)$$
$$2 = x^2 - 2x - 3x + 6$$
$$0 = x^2 - 5x + 4$$
$$0 = (x-4)(x-1)$$
$$x-4=0 \text{ or } x-1=0$$
$$x=4 \text{ or } x=1$$

53.
$$-16t^2 + 48t = 0$$
$$-16t(t-3) = 0$$
$$-16t=0 \text{ or } t-3=0$$
$$t=0 \text{ or } t=3$$
The water is at ground level at $t=0$ sec and $t=3$ sec. This agrees with the graph.

55. From the graph, the rocket will be at a height of 48 feet after $t=1$ sec and $t=3$ sec.
$$-16t^2 + 64t = 48$$
$$-16t^2 + 64t - 48 = 0$$
$$t^2 - 4t + 3 = 0$$
$$(t-3)(t-1) = 0$$
$$t-3=0 \text{ or } t-1=0$$
$$t=3 \text{ or } t=1$$
This agrees with the graph.

57. $-16t^2 + 64t = 0$
$$-16t(t-4) = 0$$
$$-16t=0 \text{ or } t-4=0$$
$$t=0 \text{ or } t=4$$
The rocket will be on the ground at $t=0$ sec and $t=4$ sec. This agrees with the graph.

59.
$$\sqrt{4-x} = x-2$$
$$\left(\sqrt{4-x}\right)^2 = (x-2)^2$$
$$4-x = x^2 - 4x + 4$$
$$x^2 - 3x = 0$$
$$x(x-3) = 0$$
$$x=0 \text{ or } x-3=0$$
$$x=0 \text{ or } x=3$$

Check:
$$\sqrt{4-0} \overset{?}{=} 0-2 \qquad \sqrt{4-3} \overset{?}{=} 3-2$$
$$\sqrt{4} \overset{?}{=} -2 \qquad\qquad \sqrt{1} \overset{?}{=} 1$$
$$2 \overset{?}{=} -2 \text{ false} \qquad\qquad 1=1 \text{ true}$$

We must discard the extraneous solution $x=0$. The only solution to the equation is $x=3$.

61.
$$\sqrt{x+5} = x+3$$
$$\left(\sqrt{x+5}\right)^2 = (x+3)^2$$
$$x+5 = x^2 + 6x + 9$$
$$x^2 + 5x + 4 = 0$$
$$(x+4)(x+1) = 0$$
$$x+4=0 \text{ or } x+1=0$$
$$x=-4 \text{ or } x=-1$$

Check:

$$\sqrt{-4+5}\overset{?}{=}-4+3 \qquad \sqrt{-1+5}\overset{?}{=}-1+3$$

$$\sqrt{1}\overset{?}{=}-1 \qquad\qquad \sqrt{4}\overset{?}{=}2$$

$$\qquad\qquad\qquad\qquad 2=2 \ \text{ true}$$

$$1=1 \ \text{ false}$$

We must discard the extraneous solution $x=-4$.
The only solution to the equation is $x=-1$.

63. a. $A(0,0)$, $B(4,16)$

    b. The intersections solve the equation
$x^2=4x$.

    c. $\qquad x^2=4x$

$$x^2-4x=0$$
$$x(x-4)=0$$
$$x=0 \ \text{ or } \ x-4=0$$
$$x=0 \ \text{ or } \ x=4$$
The results agree with the graph.

Exercises 8.6

1. $9x^2+6x+1=0$

$\qquad a=9, b=6, c=1$

3. $3x^2-9x=0$

$\qquad a=3, b=-9, c=0$

5. $\qquad x^2=4x-3$

$$x^2-4x+3=0$$
$$a=1, b=-4, c=3$$

7. $\qquad x^2=-9$

$$x^2+9=0$$
$$a=1, b=0, c=9$$

9. $\dfrac{-(-4)-\sqrt{(-4)^2-4(1)(-12)}}{2\cdot 1}$

$$=\frac{4-\sqrt{16+48}}{2}=\frac{4-\sqrt{64}}{2}=\frac{4-8}{2}=\frac{-4}{2}=-2$$

11. $\dfrac{-(-5)+\sqrt{(-5)^2-4(6)(1)}}{2(6)}$

$$=\frac{5+\sqrt{25-24}}{12}=\frac{5+\sqrt{1}}{12}=\frac{5+1}{12}=\frac{6}{12}=\frac{1}{2}$$

13. $\dfrac{-1-\sqrt{(1)^2-4(3)(-4)}}{2(3)}$

$$=\frac{-1-\sqrt{1+48}}{6}=\frac{-1-\sqrt{49}}{6}=\frac{-1-7}{6}=\frac{-8}{6}=-\frac{4}{3}$$

15. $\dfrac{-6+\sqrt{(6)^2-4(16)(-1)}}{2(16)}$

$$=\frac{-6+\sqrt{36+64}}{32}=\frac{-6+\sqrt{100}}{32}=\frac{-6+10}{32}=\frac{4}{32}=\frac{1}{8}$$

17. a. $\sqrt{2}\approx 1$

$$\frac{2-\sqrt{2}}{2}\approx\frac{2-1}{2}=\frac{1}{2}$$
Calculator:
$$\frac{2-\sqrt{2}}{2}\approx 0.293$$

    b. $\sqrt{6}\approx 2$

$$\frac{3+\sqrt{6}}{3}\approx\frac{3+2}{3}=\frac{5}{3}$$
Calculator:
$$\frac{3+\sqrt{6}}{3}\approx 1.816$$

    c. $\sqrt{6}\approx 2$

$$\frac{3\sqrt{6}}{3}\approx\frac{3(2)}{3}=2$$
Calculator:
$$\frac{3\sqrt{6}}{3}=\sqrt{6}\approx 2.449$$

19. a. $\sqrt{5}\approx 2$

$$\frac{3-\sqrt{5}}{3}\approx\frac{3-2}{3}=\frac{1}{3}$$
Calculator:
$$\frac{3-\sqrt{5}}{3}\approx 0.255$$

b. $\sqrt{10} \approx 3$

$$\frac{2\sqrt{10}}{2} \approx \frac{2(3)}{2} = 3$$

Calculator:

$$\frac{2\sqrt{10}}{2} = \sqrt{10} \approx 3.162$$

c. $\sqrt{10} \approx 3$

$$\frac{2+\sqrt{10}}{2} \approx \frac{2+3}{2} = \frac{5}{2}$$

Calculator:

$$\frac{2+\sqrt{10}}{2} \approx 2.581$$

21. a. $\sqrt{13} \approx 4$

$$\frac{-5-\sqrt{13}}{2} \approx \frac{-5-4}{2} = -\frac{9}{2}$$

Calculator:

$$\frac{-5-\sqrt{13}}{2} \approx -4.303$$

b. $\sqrt{57} \approx 8$

$$\frac{-9+\sqrt{57}}{4} \approx \frac{-9+8}{4} = -\frac{1}{4}$$

Calculator:

$$\frac{-9+\sqrt{57}}{4} \approx -0.363$$

23. a. $\sqrt{28} \approx 5$

$$\frac{-2-\sqrt{28}}{6} \approx \frac{-2-5}{6} = -\frac{7}{6}$$

Calculator:

$$\frac{-2-\sqrt{28}}{6} \approx -1.215$$

b. $\sqrt{84} \approx 9$

$$\frac{-2+\sqrt{84}}{10} \approx \frac{-2+9}{10} = \frac{7}{10}$$

Calculator:

$$\frac{-2+\sqrt{84}}{10} \approx 0.717$$

25. $4x^2 + 3x - 1 = 0$

$a = 4, b = 3, c = -1$

$$x = \frac{-3 \pm \sqrt{(3)^2 - 4(4)(-1)}}{2(4)}$$

$$= \frac{-3 \pm \sqrt{9+16}}{8}$$

$$= \frac{-3 \pm \sqrt{25}}{8}$$

$$= \frac{-3 \pm 5}{8}$$

$$x = -1 \quad \text{or} \quad x = \frac{1}{4}.$$

27. $\qquad 7x^2 = 5x + 2$

$7x^2 - 5x - 2 = 0$

$a = 7, b = -5, c = -2$

$$x = \frac{-(-5) \pm \sqrt{(-5)^2 - 4(7)(-2)}}{2(7)}$$

$$= \frac{5 \pm \sqrt{25+56}}{14}$$

$$= \frac{5 \pm \sqrt{81}}{14}$$

$$= \frac{5 \pm 9}{14}$$

$$x = -\frac{2}{7} \quad \text{or} \quad x = 1$$

29. $\qquad x = 2 - 10x^2$

$10x^2 + x - 2 = 0$

$a = 10, b = 1, c = -2$

$$x = \frac{-1 \pm \sqrt{(1)^2 - 4(10)(-2)}}{2(10)}$$

$$= \frac{-1 \pm \sqrt{1+80}}{20}$$

$$= \frac{-1 \pm \sqrt{81}}{20}$$

$$= \frac{-1 \pm 9}{20}$$

$$x = -\frac{1}{2} \quad \text{or} \quad x = \frac{2}{5}$$

31. $2x^2 - x - 1 = 0$

$a = 2, b = -1, c = -1$

$x = \dfrac{-(-1) \pm \sqrt{(-1)^2 - 4(2)(-1)}}{2(2)}$

$= \dfrac{1 \pm \sqrt{1 + 8}}{4}$

$= \dfrac{1 \pm \sqrt{9}}{4}$

$= \dfrac{1 \pm 3}{4}$

$x = -\dfrac{1}{2}$ or $x = 1$

33. $3x^2 + x - 2 = 0$

$a = 3, b = 1, c = -2$

$x = \dfrac{-1 \pm \sqrt{(1)^2 - 4(3)(-2)}}{2(3)}$

$= \dfrac{-1 \pm \sqrt{1 + 24}}{6}$

$= \dfrac{-1 \pm \sqrt{25}}{6}$

$= \dfrac{-1 \pm 5}{6}$

$x = -1$ or $x = \dfrac{2}{3}$

35. $x^2 - 2x + 2 = 0$

$a = 1, b = -2, c = 2$

$x = \dfrac{-(-2) \pm \sqrt{(-2)^2 - 4(1)(2)}}{2(1)}$

$= \dfrac{2 \pm \sqrt{4 - 8}}{2}$

$= \dfrac{2 \pm \sqrt{-4}}{2}$

The radicand is negative so the equation has no real solutions.

37. $3x^2 - 2x = 6$

$3x^2 - 2x - 6 = 0$

$a = 3, b = -2, c = -6$

$x = \dfrac{-(-2) \pm \sqrt{(-2)^2 - 4(3)(-6)}}{2(3)}$

$= \dfrac{4 \pm \sqrt{4 + 72}}{6}$

$= \dfrac{4 \pm \sqrt{76}}{6}$

$= \dfrac{4 \pm 2\sqrt{19}}{6}$

$= \dfrac{2 \pm \sqrt{19}}{3}$

$x \approx -1.120$ or $x \approx 1.786$

39. $5x^2 + 4x + 1 = 0$

$a = 5, b = 4, c = 1$

$x = \dfrac{-4 \pm \sqrt{(4)^2 - 4(5)(1)}}{2(5)}$

$= \dfrac{-4 \pm \sqrt{16 - 20}}{10}$

$= \dfrac{-4 \pm \sqrt{-4}}{10}$

The radicand is negative so the equation has no real solutions.

41. $\qquad 5x^2 = 4 - 2x$

$5x^2 + 2x - 4 = 0$

$a = 5, b = 2, c = -4$

$x = \dfrac{-2 \pm \sqrt{(2)^2 - 4(5)(-4)}}{2(5)}$

$= \dfrac{-2 \pm \sqrt{4 + 80}}{10}$

$= \dfrac{-2 \pm \sqrt{84}}{10}$

$= \dfrac{-2 \pm 2\sqrt{21}}{10}$

$= \dfrac{-1 \pm \sqrt{21}}{5}$

$x \approx -1.117$ or $x \approx 0.717$

43. $6x^2 - 3x - 2 = 0$

$a = 6, b = -3, c = -2$

$x = \dfrac{-(-3) \pm \sqrt{(-3)^2 - 4(6)(-2)}}{2(6)}$

$= \dfrac{3 \pm \sqrt{9 + 48}}{12}$

$= \dfrac{3 \pm \sqrt{57}}{12}$

$x \approx -1.137$ or $x \approx 2.637$

45. $-16t^2 + 30t + 150 = 0$

$a = -16, b = 30, c = 150$

$t = \dfrac{-30 \pm \sqrt{30^2 - 4(-16)(150)}}{2(-16)}$

$= \dfrac{-30 \pm \sqrt{10500}}{-32}$

$t \approx -2.26 \sec$ or $t \approx 4.14 \sec$

We must discard the negative solution since the time cannot be negative.

47. $-16t^2 + 15t + 150 = 0$

$a = -16, b = 15, c = 150$

$t = \dfrac{-15 \pm \sqrt{15^2 - 4(-16)(150)}}{2(-16)}$

$= \dfrac{-15 \pm \sqrt{9825}}{-32}$

$t \approx -2.63 \sec$ or $t \approx 3.57 \sec$

We must discard the negative solution since the time cannot be negative.

49. $-16t^2 + 30t + 150 = 200$

$-16t^2 + 30t - 50 = 0$

$a = -16, b = 30, c = -50$

$t = \dfrac{-30 \pm \sqrt{30^2 - 4(-16)(-50)}}{2(-16)}$

$= \dfrac{-30 \pm \sqrt{-2300}}{-32}$

Since the radicand is negative, there are no real solutions to the equation.

51. $-16t^2 + 20t + 50 = -40$

$-16^2 + 20t + 90 = 0$

$a = -16, b = 20, c = 90$

$t = \dfrac{-20 \pm \sqrt{20^2 - 4(-16)(90)}}{2(-16)}$

$= \dfrac{-20 \pm \sqrt{6160}}{-32}$

$t \approx -1.83 \sec$ or $t \approx 3.08 \sec$

We must discard the negative solution since the time cannot be negative.

53. $4x^2 = 10$

$x^2 = \dfrac{10}{4}$

$x^2 = \dfrac{5}{2}$

$\sqrt{x^2} = \sqrt{\dfrac{5}{2}}$

$|x| = \sqrt{\dfrac{5}{2}}$

$x = \pm\sqrt{\dfrac{5}{2}}$

$x \approx \pm 1.581$

55. $4x^2 - 9 = 0$

$(2x - 3)(2x + 3) = 0$

$2x + 3 = 0$ or $2x - 3 = 0$

$2x = -3$ or $2x = 3$

$x = -\dfrac{3}{2}$ or $x = \dfrac{3}{2}$

57. $x^2 + 5 = -2x$

$x^2 + 2x + 5 = 0$

$a = 1, b = 2, c = 5$

$x = \dfrac{-2 \pm \sqrt{2^2 - 4(1)(5)}}{2(1)}$

$= \dfrac{-2 \pm \sqrt{-16}}{2}$

The radicand is negative so the equation has no real solution.

59. $x^2 + 9 = 6x$

$x^2 - 6x + 9 = 0$

$(x - 3)^2 = 0$

$x - 3 = 0$

$x = 3$

61. $4x^2 = 4x + 7$

$4x^2 - 4x - 7 = 0$

$a = 4, b = -4, c = -7$

$x = \dfrac{-(-4) \pm \sqrt{(-4)^2 - 4(4)(-7)}}{2(4)}$

$= \dfrac{4 \pm \sqrt{128}}{8}$

$= \dfrac{1 \pm 2\sqrt{2}}{2}$

$x \approx -0.914$ or $x \approx 1.914$

63. $4x^2 = 12x - 9$

$4x^2 - 12x + 9 = 0$

$(2x - 3)^2 = 0$

$2x - 3 = 0$

$2x = 3$

$x = \dfrac{3}{2}$

65. We can tell the number of solutions by looking for x-intercepts. The number of solutions to a quadratic equation is the same as the number of x-intercepts.

67. First identify that $a = 2$, $b = 3$, and $c = 4$.

(Note: $2x^2 + 3x + 4 = 0$ has no real solutions)
On a TI-83 Plus, use the following keystrokes:

First solution:

[( ] [(-)] [3] [−] [2nd] [x^2] [( ]

[3] [x^2] [−] [4] [×] [2] [×]

[4] [ )] [ )] [÷] [( ] [2] [×]

[2] [ )] [ENTER]

Second solution:

[( ] [(-)] [3] [+] [2nd] [x^2] [( ]

[3] [x^2] [−] [4] [×] [2] [×]

[4] [ )] [ )] [÷] [( ] [2] [×]

[2] [ )] [ENTER]

69. Let x represent the first number and let y represent the output. The product of two consecutive numbers can be written as:

$y = x(x+1)$

Solving $y = 0$, we get:

$x(x+1) = 0$

$x = 0$ or $x = -1$
The output will be negative for values of x between -1 and 0. That is, the output will be negative on the interval $(-1, 0)$.

Exercises 8.7

1. The sum of the 6 incomes is $156,000.

$\text{Mean} = \dfrac{\$156,000}{6} = \$26,000$

Since there is an even number of values, the median is the average of the two middle values of the ordered data.

$\text{Median} = \dfrac{\$12,000 + \$13,000}{2}$

$= \dfrac{\$25,000}{2}$

$= \$12,500$

3. The sum of the 5 incomes is $130,000.

$\text{Mean} = \dfrac{\$130,000}{5} = \$26,000$

Since there is an odd number of values, the median is the center value of the ordered data.
Median = $27,500

5. Q_1 is the middle number (median) of the ordered values below the median and Q_3 is the middle number (median) of the ordered values above the median.

$Q_1 = \$10,000$

$Q_3 = \$13,000$

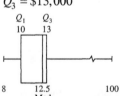

Numbers in thousands

7. Q_1 is the middle number (median) of the ordered values below the median and Q_3 is the middle number (median) of the ordered values above the median.

$$Q_1 = \frac{\$20,000 + \$27,500}{2} = \$23,750$$

$$Q_3 = \frac{\$27,500 + \$27,500}{2} = \$27,500$$

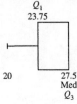

Numbers in thousands

9.

$$s = \sqrt{\frac{(8-26)^2 + (10-26)^2 + (12-26)^2 + (13-26)^2 + (13-26)^2 + (100-26)^2}{6-1}}$$

$$= \sqrt{\frac{(-18)^2 + (-16)^2 + (-14)^2 + (-13)^2 + (-13)^2 + (74)^2}{5}}$$

$$= \sqrt{\frac{324 + 256 + 196 + 169 + 169 + 5476}{5}}$$

$$= \sqrt{\frac{6590}{5}}$$

$$= \sqrt{1318}$$

≈ 36.304 thousands of dollars

That is, $s \approx \$36,304$.

11. Mean $= \$26,000$

$\sigma_x = \$3000$

$s_x \approx \$3354$

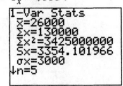

13. The average (mean) is greatly affected by extreme values while the median is not. A few high priced homes can pull the average up but have little or no effect on the median.

15. No. To calculate the standard deviation, you would need to know the actual data values.

17. Answers may vary. Student measurements may vary due to inexperience with measuring, the measurement tool that is used, etc.

19. Answers vary.

21. $3500 - 2(15) = 3500 - 30 = 3470$

$3500 + 2(15) = 3500 + 30 = 3530$

The ceramic furnace temperature is operating normally if its temperature is between $3470°F$ and $3530°F$.

23. $6.075 - 2(0.013) = 6.075 - 0.026 = 6.049$

$6.075 + 2(0.013) = 6.075 + 0.026 = 6.101$

The milling machine is operating normally if it produces components with a diameter between 6.049 cm and 6.101 cm.

25. $2.5 - 2(0.75) = 2.5 - 1.5 = 1$

$2.5 + 2(0.75) = 2.5 + 1.5 = 4$

The waiting time is considered normal if it is between 1 min and 4 min.

27. $0.30 - 1.5(0.06) = 0.30 - 0.09 = 0.21$

$0.30 + 1.5(0.06) = 0.30 + 0.09 = 0.39$

The air pollution is considered within normal bounds if there are between 0.21 ppm and 0.39 ppm.

29. $48.4 - 0.5(0.7) = 48.4 - 0.35 = 48.05$

$48.4 + 0.5(0.7) = 48.4 + 0.35 = 48.75$

The gasoline flow is considered within normal bounds if the flow rate is between 48.05 gal/min and 48.75 gal/min.

Chapter 8 Review Exercises

1. a. $2^2 = 4$, $3^2 = 9$, $4^2 = 16$

Since $2^2 + 3^2 \neq 4^2$, the numbers cannot represent the sides of a right triangle.

b. $5^2 = 25$, $17^2 = 289$, $18^2 = 324$

Since $5^2 + 17^2 \neq 18^2$, the numbers cannot represent the sides of a right triangle.

c. $8^2 = 64$, $15^2 = 225$, $17^2 = 289$

Since $8^2 + 15^2 = 17^2$, the numbers could represent the sides of a right triangle.

3. $c^2 = 6^2 + 24^2$

$c^2 = 36 + 576$

$c^2 = 612$

$c = \sqrt{612} \approx 24.74$

The ladder would need to be about 24.74 feet long.

5. a. $\sqrt{60} = \sqrt{4 \cdot 15} = \sqrt{4} \cdot \sqrt{15} = 2\sqrt{15}$

b. $\sqrt{63} = \sqrt{9 \cdot 7} = \sqrt{9} \cdot \sqrt{7} = 3\sqrt{7}$

c. $\sqrt{54} = \sqrt{9 \cdot 6} = \sqrt{9} \cdot \sqrt{6} = 3\sqrt{6}$

7. a. $144^{1/2} = \sqrt{144} = 12$

b. $144^{-1} = \dfrac{1}{144}$

c. $144^0 = 1$

d. $144^{0.5} = 144^{1/2} = \sqrt{144} = 12$

9. a. $(0.36)^{1/2} = \sqrt{0.36} = 0.6$

b. $(0.36)^{-1} = \left(\dfrac{36}{100}\right)^{-1} = \dfrac{100}{36} = \dfrac{25}{9}$

c. $(0.36)^{0.5} = (0.36)^{1/2} = \sqrt{0.36} = 0.6$

d. $(0.36)^0 = 1$

11. a. Sides:

$\sqrt{(-2-0)^2 + (3-0)^2} = \sqrt{13}$

$\sqrt{(-2-1)^2 + (3-5)^2} = \sqrt{13}$

$\sqrt{(1-3)^2 + (5-2)^2} = \sqrt{13}$

$\sqrt{(3-0)^2 + (2-0)^2} = \sqrt{13}$

Diagonals:

$\sqrt{(1-0)^2 + (5-0)^2} = \sqrt{26}$

$\sqrt{(-2-3)^2 + (3-2)^2} = \sqrt{26}$

The shape is a square.

b. Sides:

$\sqrt{(-4-0)^2 + (2-0)^2} = \sqrt{20} = 2\sqrt{5}$

$\sqrt{(-6-(-4))^2 + (-2-2)^2} = \sqrt{20} = 2\sqrt{5}$

$\sqrt{(-2-(-6))^2 + (-4-(-2))^2} = \sqrt{20} = 2\sqrt{5}$

$\sqrt{(-2-0)^2 + (-4-0)^2} = \sqrt{20} = 2\sqrt{5}$

Diagonals:

$\sqrt{(-6-0)^2 + (-2-0)^2} = \sqrt{40} = 2\sqrt{10}$

$\sqrt{(-2-(-4))^2 + (-4-2)^2} = \sqrt{40} = 2\sqrt{10}$

The shape is a square.

c. Sides:

$$\sqrt{(-1-0)^2+(3-0)^2}=\sqrt{10}$$

$$\sqrt{(-1-3)^2+(3-4)^2}=\sqrt{17}$$

$$\sqrt{(3-4)^2+(4-1)^2}=\sqrt{10}$$

$$\sqrt{(4-0)^2+(1-0)^2}=\sqrt{17}$$

Diagonals:

$$\sqrt{(3-0)^2+(4-0)^2}=\sqrt{25}=5$$

$$\sqrt{(-1-4)^2+(3-1)^2}=\sqrt{29}=3\sqrt{3}$$

The shape is not a rectangle.

13. a. $\sqrt{25x^2y^4}=\sqrt{25}\sqrt{x^2}\sqrt{(y^2)^2}=5xy^2$

b. $\sqrt{169x^6y^2}=\sqrt{169}\sqrt{(x^3)^2}\sqrt{y^2}$
$$=13x^3y$$

c. $\sqrt{2.25a^3}=\sqrt{2.25}\sqrt{a^2a}=1.5a\sqrt{a}$

d. $\sqrt{0.64b^5}=\sqrt{0.64}\sqrt{(b^2)^2b}=0.8b^2\sqrt{b}$

e. $\sqrt{\dfrac{80x^3}{5x}}=\sqrt{16x^2}=\sqrt{16}\sqrt{x^2}=4x$

f. $\sqrt{\dfrac{3a^4}{27b^6}}=\sqrt{\dfrac{a^4}{9b^6}}=\dfrac{\sqrt{(a^2)^2}}{\sqrt{9}\sqrt{(b^3)^2}}=\dfrac{a^2}{3b^3}$

g. $\sqrt{\dfrac{192a^6}{3}}=\sqrt{64a^6}=\sqrt{64}\sqrt{(a^3)^2}=8a^3$

h. $\sqrt{\dfrac{121}{49b^4}}=\dfrac{\sqrt{121}}{\sqrt{49}\sqrt{(b^2)^2}}=\dfrac{11}{7b^2}$

15. The results are almost identical to the f-stops.

n	$\left(\sqrt{2}\right)^n$
1	$\left(\sqrt{2}\right)^1\approx1.41$
2	$\left(\sqrt{2}\right)^2=2$
3	$\left(\sqrt{2}\right)^3\approx2.83$
4	$\left(\sqrt{2}\right)^4=4$
5	$\left(\sqrt{2}\right)^5\approx5.66$
6	$\left(\sqrt{2}\right)^6=8$
7	$\left(\sqrt{2}\right)^7\approx11.3$
8	$\left(\sqrt{2}\right)^8=16$

17. a.
$$\sqrt{7x-3}=5$$
$$\left(\sqrt{7x-3}\right)^2=5^2$$
$$7x-3=25$$
$$7x=28$$
$$x=4$$

Check:

$$\sqrt{7(4)-3}\overset{?}{=}5$$

$$\sqrt{25}\overset{?}{=}5$$

$$5=5 \text{ true}$$

The radical expression is defined when $7x-3\geq0$. That is,

$$7x-3\geq0$$
$$7x\geq3$$
$$x\geq\dfrac{3}{7}$$

b.
$$\sqrt{2-x}=x-2$$
$$\left(\sqrt{2-x}\right)^2=(x-2)^2$$
$$2-x=x^2-4x+4$$
$$0=x^2-3x+2$$
$$0=(x-2)(x-1)$$
$$x-2=0 \text{ or } x-1=0$$
$$x=2 \text{ or } x=1$$

Check:

$$\sqrt{2-2}\overset{?}{=}2-2 \qquad \sqrt{2-1}\overset{?}{=}1-2$$

$$\sqrt{0}\overset{?}{=}0 \qquad\qquad \sqrt{1}\overset{?}{=}-1$$

$$0=0 \text{ true} \qquad\qquad \overset{?}{1=-1} \text{ false}$$

We must discard the extraneous solution $x-1$. The only solution is $x=2$.
The radical expression is defined when $2-x \ge 0$, or $x \le 2$.

c.　　　$\sqrt{4x-3}=7$

$$\left(\sqrt{4x-3}\right)^2 = 7^2$$

$$4x-3=49$$

$$4x=52$$

$$x=13$$

Check:

$$\sqrt{4(13)-3}\overset{?}{=}7$$

$$\sqrt{49}\overset{?}{=}7$$

$$7=7 \text{ true}$$

The radical expression is defined when $4x-3 \ge 0$. That is,

$$4x-3 \ge 0$$

$$4x \ge 3$$

$$x \ge \frac{3}{4}$$

19. a.　$d = \sqrt{\dfrac{3h}{8}} = \sqrt{\dfrac{3(20)}{8}} = \sqrt{7.5} \approx 2.7$

From a height of 20 feet on the moon, you can see a distance of about 2.7 miles.

b.　　　$4 = \sqrt{\dfrac{3h}{8}}$

$$4^2 = \left(\sqrt{\dfrac{3h}{8}}\right)^2$$

$$16 = \dfrac{3h}{8}$$

$$128 = 3h$$

$$\dfrac{128}{3} = h \quad \text{or} \quad h = 42\tfrac{2}{3}$$

To see a distance of 4 miles on the moon, you would need to at a height of $42\tfrac{2}{3}$ feet.

21.　　$\sqrt{x-1}=2$

$$\left(\sqrt{x-1}\right)^2 = 2^2$$

$$x-1=4$$

$$x=5$$

The solution can be found from the graph by locating the intersection point of the graph of $y=\sqrt{x-1}$ and the graph of $y=2$. The x-coordinate of the intersection point is the solution.

23. From the graph, $x^2-2x-8=0$ when $x=-2$ and $x=4$.

25.　　$x^2 = \dfrac{16}{144}$

$$\sqrt{x^2} = \sqrt{\dfrac{16}{144}}$$

$$|x| = \dfrac{1}{3}$$

$$x = \pm\dfrac{1}{3}$$

27.　　$x^2-5x+4=0$

$$(x-4)(x-1)=0$$

$$x-4=0 \text{ or } x-1=0$$

$$x=4 \text{ or } x=1$$

29.　$2x^2+5x-6=0$

$$a=2, b=5, c=-6$$

$$x = \dfrac{-5 \pm \sqrt{5^2-4(2)(-6)}}{2(2)}$$

$$= \dfrac{-5 \pm \sqrt{73}}{4}$$

$$x \approx -3.386 \text{ or } x \approx 0.886$$

31.　　$x^2+3x-18=0$

$$(x+6)(x-3)=0$$

$$x+6=0 \text{ or } x-3=0$$

$$x=-6 \text{ or } x=3$$

174

33.
$$3x^2 = 4x + 7$$
$$3x^2 - 4x - 7 = 0$$
$$(3x - 7)(x + 1) = 0$$
$$3x - 7 = 0 \text{ or } x + 1 = 0$$
$$3x = 7 \text{ or } x = -1$$
$$x = \frac{7}{3} \text{ or } x = -1$$

35. $2x^2 + 3x + 5 = 0$
$$a = 2, b = 3, c = 5$$
$$x = \frac{-3 \pm \sqrt{3^2 - 4(2)(5)}}{2(2)}$$
$$= \frac{-3 \pm \sqrt{-31}}{4}$$

The radicand is negative so the equation has no real solutions.

37. $x^2 - 6x + 9 = 0$
$$(x - 3)^2 = 0$$
$$x - 3 = 0$$
$$x = 3$$

39. a. $A = \pi r^2$
$$\frac{A}{\pi} = r^2$$
$$\sqrt{\frac{A}{\pi}} = \sqrt{r^2}$$
$$\sqrt{\frac{A}{\pi}} = r \text{ or } r = \sqrt{\frac{A}{\pi}}$$

    b. $p = \frac{1}{2}dv^2$
$$2p = dv^2$$
$$\frac{2p}{d} = v^2$$
$$\sqrt{\frac{2p}{d}} = \sqrt{v^2}$$
$$\sqrt{\frac{2p}{d}} = v \text{ or } v = \sqrt{\frac{2p}{d}}$$

    c. $S = 4\pi r^2$
$$\frac{S}{4\pi} = r^2$$
$$\sqrt{\frac{S}{4\pi}} = \sqrt{r^2}$$
$$\frac{1}{2}\sqrt{\frac{S}{\pi}} = r \text{ or } r = \frac{1}{2}\sqrt{\frac{S}{\pi}}$$

    d. $h = \frac{v^2}{2g}$
$$2gh = v^2$$
$$\sqrt{2gh} = \sqrt{v^2}$$
$$\sqrt{2gh} = v \text{ or } v = \sqrt{2gh}$$

41. a. <u>Men:</u>
$$\text{range} = \text{max} - \text{min}$$
$$= 39.17 - 34.42$$
$$= 4.75 \text{ sec}$$

    <u>Women:</u>
$$\text{range} = \text{max} - \text{min}$$
$$= 42.76 - 37.30$$
$$= 5.46 \text{ sec}$$

    b. <u>Men:</u>
The median is the average of the two middle values.
$$\text{median} = \frac{36.45 + 37.14}{2} = 36.795 \text{ sec}$$

The quartiles, Q_1 and Q_3 are the medians of the lower and upper half of the ordered data, respectively.
$$Q_3 = \frac{38.03 + 38.19}{2} = 38.11 \text{ sec}$$
$$Q_1 = \frac{35.59 + 36.33}{2} = 35.96 \text{ sec}$$

    <u>Women:</u>
The median is the average of the two middle values.
$$\text{median} = \frac{40.33 + 39.25}{2} = 39.79 \text{ sec}$$

The quartiles, Q_1 and Q_3 are the medians of the lower and upper half of the ordered data, respectively.

$$Q_3 = \frac{41.78 + 41.02}{2} = 41.4 \text{ sec}$$

$$Q_1 = \frac{38.21 + 39.10}{2} = 38.655 \text{ sec}$$

The boxplots are given below.

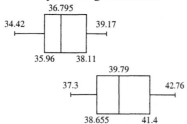

Chapter 8 Test

1. a. $3^2 = 9$, $4^2 = 16$, $5^2 = 25$

Since $3^2 + 4^2 = 5^2$, the numbers could represent the sides of a right triangle.

b. $1^2 = 1$, $2^2 = 4$, $3^2 = 9$

Since $1^2 + 2^2 \neq 3^2$, the numbers cannot represent the sides of a right triangle.

c. $1^2 = 1$, $\left(\sqrt{3}\right)^2 = 3$, $2^2 = 4$

Since $1^2 + \left(\sqrt{3}\right)^2 = 2^3$, the numbers could represent the sides of a right triangle.

d. $\left(\sqrt{2}\right)^2 = 2$, $\left(\sqrt{2}\right)^2 = 2$, $2^2 = 4$

Since $\left(\sqrt{2}\right)^2 + \left(\sqrt{2}\right)^2 = 2^2$, the numbers could represent the sides of a right triangle.

e. $\left(\sqrt{3}\right)^2 = 3$, $\left(\sqrt{4}\right)^2 = 4$, $\left(\sqrt{5}\right)^2 = 5$

Since $\left(\sqrt{3}\right)^2 + \left(\sqrt{4}\right)^2 \neq \left(\sqrt{5}\right)^2$, the numbers cannot represent the sides of a right triangle.

2. a. $\sqrt{45} = \sqrt{9 \cdot 5} = \sqrt{9} \cdot \sqrt{5} = 3\sqrt{5}$

b. $\sqrt{44} = \sqrt{4 \cdot 11} = \sqrt{4} \cdot \sqrt{11} = 2\sqrt{11}$

3. $c^2 = 5^2 + 20^2$

$c^2 = 25 + 400$

$c^2 = 425$

$c = \sqrt{425} \approx 20.6$

The ladder would need to be about 20.6 feet long.

4. a. $\sqrt{5} \cdot \sqrt{20} = \sqrt{5 \cdot 20} = \sqrt{100} = 10$

b. $\sqrt{3} \cdot \sqrt{27} = \sqrt{3 \cdot 27} = \sqrt{81} = 9$

c. $\left(3\sqrt{6}\right)^2 = 3^2 \cdot \left(\sqrt{6}\right)^2 = 9 \cdot 6 = 54$

d. $\left(2\sqrt{7}\right)^2 = 2^2 \left(\sqrt{7}\right)^2 = 4 \cdot 7 = 28$

e. $\sqrt{36x^2 y} = \sqrt{36x^2} \cdot \sqrt{y} = 6x\sqrt{y}$

f. $\sqrt{0.81x^4 y^3} = \sqrt{0.81x^4 y^2} \cdot \sqrt{y} = 0.9x^2 y\sqrt{y}$

g. $\sqrt{\dfrac{147}{3}} = \sqrt{49} = 7$

h. $\dfrac{\sqrt{18}}{\sqrt{32}} = \sqrt{\dfrac{18}{32}} = \sqrt{\dfrac{9}{16}} = \dfrac{\sqrt{9}}{\sqrt{16}} = \dfrac{3}{4}$

i. $\sqrt{\dfrac{a^5 b^2}{a}} = \sqrt{a^4 b^2} = a^2 b$

5. a. $a^2 = \left(\sqrt{13}\right)^2 + 6^2$

$a^2 = 13 + 36$

$a^2 = 49$

$a = \sqrt{49}$

$a = 7$

b. $\dfrac{b}{18} = \dfrac{7}{6}$ $\qquad$ $\dfrac{c}{18} = \dfrac{\sqrt{13}}{6}$

$b = \dfrac{18 \cdot 7}{6}$ $\qquad$ $c = \dfrac{18\sqrt{13}}{6}$

$b = 21$ $\qquad$ $c = 3\sqrt{13}$

$\qquad\qquad\qquad c \approx 10.8$

c.

$$\frac{d}{10} = \frac{6}{\sqrt{13}} \qquad \frac{e}{10} = \frac{7}{\sqrt{13}}$$

$$d = \frac{10 \cdot 6}{\sqrt{13}} \qquad e = \frac{10 \cdot 7}{\sqrt{13}}$$

$$d = \frac{60}{\sqrt{13}} \qquad e = \frac{70}{\sqrt{13}}$$

$$d \approx 16.6 \qquad e \approx 19.4$$

6. From the graph, $x^2 - 3 = 1$ when $x = -2$ and $x = 2$.

7. a.

$$x^2 = \frac{36}{121}$$

$$\sqrt{x^2} = \sqrt{\frac{36}{121}}$$

$$|x| = \frac{6}{11}$$

$$x = \pm\frac{6}{11}$$

b.

$$\sqrt{3-x} = x+3$$

$$\left(\sqrt{3-x}\right)^2 = (x+3)^2$$

$$3 - x = x^2 + 6x + 9$$

$$0 = x^2 + 7x + 6$$

$$0 = (x+6)(x+1)$$

$$x + 6 = 0 \quad \text{or} \quad x + 1 = 0$$

$$x = -6 \quad \text{or} \quad x = -1$$

Check:

$$\sqrt{3-(-6)} \overset{?}{=} -6+3 \qquad \sqrt{3-(-1)} \overset{?}{=} -1+3$$

$$\sqrt{9} \overset{?}{=} -3 \qquad\qquad \sqrt{4} \overset{?}{=} 2$$

$$3 \overset{?}{=} -3 \text{ false} \qquad\qquad 2 = 2 \text{ true}$$

We need to discard the extraneous solution $x = -6$. The only solution is $x = -1$. The radical expression is defined when $3 - x \geq 0$, or $x \leq 3$.

c.

$$x^2 + x - 2 = 0$$

$$(x+2)(x-1) = 0$$

$$x + 2 = 0 \quad \text{or} \quad x - 1 = 0$$

$$x = -2 \quad \text{or} \quad x = 1$$

d.

$$\sqrt{5x-6} = 12$$

$$\left(\sqrt{5x-6}\right)^2 = 12^2$$

$$5x - 6 = 144$$

$$5x = 150$$

$$x = 30$$

The radical expression is defined when $5x - 6 \geq 0$. That is,

$$5x - 6 \geq 0$$

$$5x \geq 6$$

$$x \geq \frac{6}{5}$$

e.

$$2x^2 = 8 - 15x$$

$$2x^2 + 15x - 8 = 0$$

$$(2x-1)(x+8) = 0$$

$$2x - 1 = 0 \quad \text{or} \quad x + 8 = 0$$

$$2x = 1 \quad \text{or} \quad x = -8$$

$$x = \frac{1}{2} \quad \text{or} \quad x = -8$$

f.

$$8x^2 - 5x = 4$$

$$8x^2 - 5x - 4 = 0$$

$$a = 8, b = -5, c = -4$$

$$x = \frac{-(-5) \pm \sqrt{(-5)^2 - 4(8)(-4)}}{2(8)}$$

$$= \frac{5 \pm \sqrt{153}}{16}$$

$$x \approx -0.461 \quad \text{or} \quad x \approx 1.086$$

g.

$$(3x+8)(3x-8) = 0$$

$$3x + 8 = 0 \quad \text{or} \quad 3x - 8 = 0$$

$$3x = -8 \quad \text{or} \quad 3x = 8$$

$$x = -\frac{8}{3} \quad \text{or} \quad x = \frac{8}{3}$$

h.

$$4x^2 + 8 = 0$$

$$x^2 + 2 = 0$$

$$x^2 = -2$$

This equation has no real solution. In the real number system, $x^2 \geq 0$.

i. $x^2 - 6x + 9 = 0$

$(x-3)^2 = 0$

$x - 3 = 0$

$x = 3$

j. $4x^2 - 25 = 0$

$(2x-5)(2x+5) = 0$

$2x - 5 = 0$ or $2x + 5 = 0$

$2x = 5$ or $2x = -5$

$x = \dfrac{5}{2}$ or $x = -\dfrac{5}{2}$

8. a. $V_e = \sqrt{\dfrac{2GM}{R}}$

$V_e^2 = \left(\sqrt{\dfrac{2GM}{R}}\right)^2$

$V_e^2 = \dfrac{2GM}{R}$

$R \cdot V_e^2 = 2GM$

$R = \dfrac{2GM}{V_e^2}$

b. $E = \dfrac{1}{2}mv^2$

$2E = mv^2$

$\dfrac{2E}{m} = v^2$

$\sqrt{\dfrac{2E}{m}} = v$ or $v = \sqrt{\dfrac{2E}{m}}$

c. $E = \dfrac{kH^2}{8\pi}$

$8\pi E = kH^2$

$\dfrac{8\pi E}{k} = H^2$

$\sqrt{\dfrac{8\pi E}{k}} = H$

$2\sqrt{\dfrac{2\pi E}{k}} = H$ or $H = 2\sqrt{\dfrac{2\pi E}{k}}$

9. a. $s = 8.6\sqrt{t}$

$= 8.6\sqrt{36}$

$= (8.6)(6)$

$= 51.6$

If the tire pressure is 36 pounds per square inch, the tire will hydroplane when the speed is 51.6 miles per hour.

b. $s = 8.6\sqrt{t}$

$= 8.6\sqrt{100}$

$= (8.6)(10)$

$= 86$

If the tire pressure is 100 pounds per square inch, the tire will hydroplane when the speed is 86 miles per hour.

c. $s = 8.6\sqrt{t}$

$120 = 8.6\sqrt{t}$

$\dfrac{120}{8.6} = \sqrt{t}$

$\left(\dfrac{120}{8.6}\right)^2 = t$

$194.7 \approx t$

A plane landing at a speed of 120 miles per hour will hydroplane if the tire pressure is less than 194.7 pounds per square inch.

10. a. $A = x^2$

$50 = x^2$

$\sqrt{50} = x$

$5\sqrt{2} = x$

$7.071 \approx x$

Each side of the square is roughly 7.071 ft.

b. $A = \pi r^2$

$50 = \pi r^2$

$\dfrac{50}{\pi} = r^2$

$\sqrt{\dfrac{50}{\pi}} = r$

$3.989 \approx r$

The radius of the circle is roughly 3.989 ft.

c.
$$A = \frac{x^2\sqrt{3}}{4}$$
$$50 = \frac{x^2\sqrt{3}}{4}$$
$$200 = x^2\sqrt{3}$$
$$\frac{200}{\sqrt{3}} = x^2$$
$$\sqrt{\frac{200}{\sqrt{3}}} = x$$
$$10.746 \approx x$$

The triangle has sides measuring roughly 10.746 ft.

11. a.

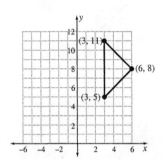

b.
$$\sqrt{(3-6)^2 + (5-8)^2} = \sqrt{18} \approx 4.24$$
$$\sqrt{(6-3)^2 + (8-11)^2} = \sqrt{18} \approx 4.24$$
$$\sqrt{(3-3)^2 + (11-5)^2} = \sqrt{36} = 6$$

c. $\dfrac{5-8}{3-6} = 1$, $\dfrac{8-11}{6-3} = -1$, $\dfrac{11-5}{3-3} =$ undefined

The two non-vertical sides are perpendicular because their slopes are opposite-reciprocals (i.e. the product of their slopes $= -1$).

d. The two non-vertical sides have the same length and meet at a right angle (since they are perpendicular). Thus, this is an isosceles right triangle.

12. a. Mean $= \dfrac{120+110+105+110+105+110}{6} = \dfrac{660}{6} = 110$ calories

$$s = \sqrt{\frac{(120-110)^2 + (110-110)^2 + (105-110)^2 + (110-110)^2 + (105-110)^2 + (110-110)^2}{5}}$$
$$= \sqrt{\frac{10^2 + 0^2 + (-5)^2 + 0^2 + (-5)^2 + 0^2}{5}}$$
$$= \sqrt{\frac{100 + 25 + 25}{5}}$$
$$= \sqrt{\frac{150}{5}}$$
$$= \sqrt{30}$$
$$\approx 5.5 \text{ calories}$$

b. The median is the middle value of the ordered data. Since there are an even number of values, the median will be the average of the two middle values.

$$\text{Median} = \frac{145 + 145}{2} = 145 \text{ mg}$$

The quartiles, Q_1 and Q_3 are the medians of the lower and upper half of the ordered data, respectively.

$Q_1 = 75$ mg

$Q_3 = 181$ mg

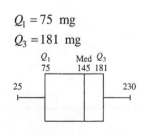

Cumulative Review of Chapters 1 to 8

1. $12 \text{ hr} \div 1.25 \text{ hr} = 9$
9 procedures can be scheduled in a 12-hour day.

3. $3(\$15) + 2(-\$20) - (\$10) - (-\$25)$
$= \$45 - \$20 - \$10 + \$25$
$= \$20$

5. $\dfrac{3x^3 y^2}{27xy^3} = \dfrac{3}{27} x^{3-1} y^{2-3} = \dfrac{1}{9} x^2 y^{-1} = \dfrac{x^2}{9y}$

7. $2 - 3x = 26$
$-3x = 24$
$x = -8$

9. $3(x+4) = 2(1-x)$
$3x + 12 = 2 - 2x$
$5x = -10$
$x = -2$

11. $2x - 1 \le 2 - x$
$3x \le 3$
$x \le 1$

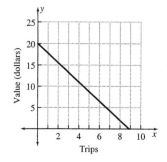

13.

x	$f(x) = 3x - 2x^2$
-1	$3(-1) - 2(-1)^2 = -5$
0	$3(0) - 2(0)^2 = 0$
1	$3(1) - 2(1)^2 = 1$
2	$3(2) - 2(2)^2 = -2$

15. $m = \dfrac{4}{3}$, $b = -2$
$y = mx + b$
$y = \dfrac{4}{3}x - 2$

17. $y = 20 - 2.25x$

Trips, x	Remaining value (\$), y
0	20
2	$20 - 2(2.25) = 15.50$
4	$20 - 4(2.25) = 11.00$
6	$20 - 6(2.25) = 6.50$
8	$20 - 8(2.25) = 2.00$
10	$20 - 10(2.25) = -2.50$

19. $\dfrac{x}{52} = \dfrac{125}{100}$

$x = 52\left(\dfrac{125}{100}\right)$

$x = 65$

21. $\dfrac{x}{100} = \dfrac{33}{88}$

$x = 100\left(\dfrac{33}{88}\right)$

$x = 37.5\%$

23. $2 \text{ sec} \cdot \dfrac{100 \text{ miles}}{1 \text{ hr}} \cdot \dfrac{1 \text{ hr}}{3600 \text{ sec}} \cdot \dfrac{5280 \text{ ft}}{1 \text{ mile}} = 293\frac{1}{3} \text{ ft}$
Sneeze bacteria travels $293\frac{1}{3}$ feet in 2 seconds.

25. a. $(x+3)(2x-1) = 2x^2 - x + 6x - 3$
$= 2x^2 + 5x - 3$

   b. $x(x+1)(3x-1) = x(3x^2 - x + 3x - 1)$
$= x(3x^2 + 2x - 1)$
$= 3x^3 + 2x^2 - x$

27. a. $\dfrac{1.496 \times 10^8}{1.3914 \times 10^6} = \dfrac{1.496}{1.3914} \times 10^{8-6} \approx 1.075 \times 10^2$

b. $\dfrac{384{,}000}{3480} = \dfrac{3.84 \times 10^5}{3.48 \times 10^3}$

$= \dfrac{3.84}{3.48} \times 10^{5-3}$

$\approx 1.103 \times 10^2$

c. The ratios are similar, about 100 to 1.

29. $7x - 8y = -25$ $\qquad 7x - 8y = -25$

$3x + 4y = -7$ $\times 2 \qquad \underline{6x + 8y = -14}$

$\qquad\qquad\qquad\qquad 13x = -39$

$\qquad\qquad\qquad\qquad\quad x = -3$

Substitute $x = -3$ into the second equation and solve for y.

$3(-3) + 4y = -7$

$-9 + 4y = -7$

$4y = 2$

$y = \dfrac{1}{2}$

The ordered pair solution is $\left(-3, \dfrac{1}{2}\right)$.

31. Let x be the length of the short leg. Then the longer leg has length $x + 3$.

$x^2 + (x+3)^2 = 15^2$

$x^2 + x^2 + 6x + 9 = 225$

$2x^2 + 6x - 216 = 0$

$x^2 + 3x - 108 = 0$

$(x - 9)(x + 12) = 0$

$x - 9 = 0$ or $x + 12 = 0$

$x = 9$ or $x = -12$

Since lengths are not negative, the shorter leg must have a length of 9 units. The longer leg has a length of $9 + 3 = 12$ units.

33. a. $y = 2x^2 - 5x - 3$

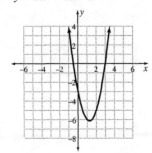

b. From the graph, $2x^2 - 5x - 3 = 0$ when

$x = -\dfrac{1}{2}$ and $x = 3$.

c. $2x^2 - 5x - 3 = 0$

$(2x + 1)(x - 3) = 0$

$2x + 1 = 0$ or $x - 3 = 0$

$2x = -1$ or $x = 3$

$x = -\dfrac{1}{2}$ or $x = 3$

d. $2x^2 - 5x - 3 = 0$

$a = 2, b = -5, c = -3$

$x = \dfrac{-(-5) \pm \sqrt{(-5)^2 - 4(2)(-3)}}{2(2)}$

$= \dfrac{5 \pm \sqrt{49}}{4}$

$= \dfrac{5 \pm 7}{4}$

$x = \dfrac{5 - 7}{4} = -\dfrac{1}{2}$ or $x = \dfrac{5 + 7}{4} = 3$

35. a. Mean $= \dfrac{5+5+3+4+5+5+6}{7} = \dfrac{33}{7} \approx 4.7$ g

There is an even number of values so the median is the middle value of the ordered data.

$3, 4, 5, 5, 5, 5, 6$

The median is 5 g.

The mode is 5 g, the most frequent value.

b. The quartiles, Q_1 and Q_3 are the medians of the lower and upper half of the ordered data, respectively.

$Q_1 = 4$ and $Q_3 = 5$

Using min, Q_1, median, Q_3, max, the boxplot is:

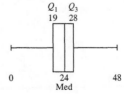

c. <u>Range:</u>

$\text{max} - \text{min}$

$= 348 - 237$

$= 111 \text{ calories}$

<u>Mean:</u>

$$\frac{312 + 295 + 237 + 289 + 302 + 303 + 348}{7}$$

$$= \frac{2086}{7} = 298 \text{ calories}$$

<u>Sample standard deviation:</u>

$$s_x = \sqrt{\frac{14^2 + (-3)^2 + (-61)^2 + (-9)^2 + 4^2 + 5^2 + 50^2}{6}}$$

$$= \sqrt{\frac{6548}{6}}$$

$$\approx 33 \text{ calories}$$

Chapter 9

Exercises 9.1

1. $\mathbb{R}, x \neq -1$

3. $\mathbb{R}, x \neq \dfrac{1}{2}$

5. $\mathbb{R}, x \neq -\dfrac{1}{3}$

7. $\dfrac{5}{x^2+4x+3} = \dfrac{5}{(x+1)(x+3)}$;
$\mathbb{R}, x \neq -3, x \neq -1$

9. y approaches $-\infty$ from the left, $+\infty$ from the right.

11. y approaches $+\infty$ from the left, $-\infty$ from the right.

13. A rate of zero, $r = 0$, is interpreted as no motion.

15. No; the numerator will never equal zero.

17. Solving for r in $D = rt$, we get $r = \dfrac{D}{t}$.

a. $r = \dfrac{60 \text{ mi}}{2 \text{ hr}} = 30 \text{ mph}$

b. $r = \dfrac{60 \text{ mi}}{1 \text{ hr}} = 60 \text{ mph}$

c. $r = \dfrac{60 \text{ mi}}{{}^{5}\!/_{6} \text{ hr}} = \dfrac{60 \text{ mi}}{1} \cdot \dfrac{6}{5 \text{ hr}} = 72 \text{ mph}$

d. $r = \dfrac{60 \text{ mi}}{{}^{3}\!/_{4} \text{ hr}} = \dfrac{60 \text{ mi}}{1} \cdot \dfrac{4}{3 \text{ hr}} = 80 \text{ mph}$

e. $r = \dfrac{60 \text{ mi}}{{}^{2}\!/_{3} \text{ hr}} = \dfrac{60 \text{ mi}}{1} \cdot \dfrac{3}{2 \text{ hr}} = 90 \text{ mph}$

f. $r = \dfrac{60 \text{ mi}}{{}^{1}\!/_{2} \text{ hr}} = \dfrac{60 \text{ mi}}{1} \cdot \dfrac{2}{1 \text{ hr}} = 120 \text{ mph}$

19. a.

Width	Length	Area	Perimeter
1	30	30	62
2	15	30	34
3	10	30	26
4	7.5	30	23
5	6	30	22
6	5	30	22
7.5	4	30	23
10	3	30	26
15	2	30	34
30	1	30	62

b.

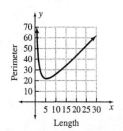

c. From the graph above, the point on the graph describing the smallest perimeter is approximately $(5.5, 21.9)$.

d. Let $x = \text{width}$. Since $A = l \cdot w$ and the area of the rectangles is 30 in^2, we can write
$30 = l \cdot x$
$l = \dfrac{30}{x}$.
The perimeter of a rectangle is $P = 2w + 2l$, so we can write $P = 2x + 2\left(\dfrac{30}{x}\right)$.

21. For $x \neq 0$, $y = \dfrac{1000 \times 10^9}{x}$,
$y = \dfrac{1000 \times 10^9}{365x}$

23. $y = \dfrac{1000 \times 10^9}{365(74 \times 10^6)} = \dfrac{1000 \times 10^9}{2.701 \times 10^{10}} \approx 37 \text{ yr}$

25. a. y is the number of terms the aid will last.

b. $y = \dfrac{90}{12} = \dfrac{15}{2} = 7\dfrac{1}{2}$.

       The aid will last $7\dfrac{1}{2}$ months.

27. $r = \dfrac{1200 \text{ miles}}{9 \text{ wks}} \cdot \dfrac{1 \text{ wk}}{7 \text{ days}} \cdot \dfrac{1 \text{ day}}{16 \text{ hrs}} \approx 1.2$ mph

29. a.

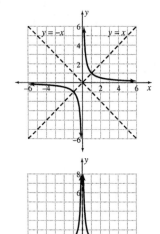

b.   All values of x give a positive number for y.

c.   Yes; $y = \dfrac{1}{x}$ is equivalent to $x = \dfrac{1}{y}$.

d. $y = \dfrac{1}{x}$; as x gets larger, x^2 becomes larger than x.

Exercises 9.2

1. $\dfrac{2ab}{6a^2 b} = \dfrac{2 \cdot a \cdot b}{2 \cdot 3 \cdot a \cdot a \cdot b} = \dfrac{1}{3a}$

3. $\dfrac{12cd^2}{8c^2 d} = \dfrac{3 \cdot 4 \cdot c \cdot d \cdot d}{2 \cdot 4 \cdot c \cdot c \cdot d} = \dfrac{3d}{2c}$

5. $\dfrac{(x-2)(x+3)}{x+3} = x-2$

7. $\dfrac{2-x}{(x+2)(x-2)} = \dfrac{-1(x-2)}{(x+2)(x-2)} = \dfrac{-1}{(x+2)}$

9. $\dfrac{3ab + 3ac}{5b^2 + 5bc} = \dfrac{3a(b+c)}{5b(b+c)} = \dfrac{3a}{5b}$

11. $\dfrac{4x^2 + 8x}{2x^2 - 4x} = \dfrac{4x(x+2)}{2x(x-2)} = \dfrac{2(x+2)}{x-2}$

13. $\dfrac{x^2 - 4}{x^2 + 5x + 6} = \dfrac{(x-2)(x+2)}{(x+3)(x+2)} = \dfrac{x-2}{x+3}$

15. $\dfrac{x^2 + x - 6}{x^2 - 2x} = \dfrac{(x+3)(x-2)}{x(x-2)} = \dfrac{x+3}{x}$

17. $\dfrac{x-3}{6-2x} = \dfrac{x-3}{-2(x-3)} = -\dfrac{1}{2}$

19. $\dfrac{x^2 - 5x + 6}{x^2 - 9} = \dfrac{(x-2)(x-3)}{(x+3)(x-3)} = \dfrac{x-2}{x+3}$

21. $\dfrac{6}{9} = \dfrac{6 \cdot 5}{9 \cdot 5} = \dfrac{30}{45}$

23. $\dfrac{24x}{3x^2} = \dfrac{8 \cdot 3x}{x \cdot 3x} = \dfrac{8}{x}$

25. $\dfrac{b}{2a} = \dfrac{b \cdot 5a}{2a \cdot 5a} = \dfrac{5ab}{10a^2}$

27. $\dfrac{2x+4}{2x-6} = \dfrac{2(x+2)}{2(x-3)} = \dfrac{x+2}{x-3}$

29. $\dfrac{a^2 + 2ab + b^2}{a^2 - b^2} = \dfrac{(a+b)(a+b)}{(a-b)(a+b)} = \dfrac{a+b}{a-b}$

31. $\dfrac{108 \text{ m}^2}{6 \text{ m}} = \dfrac{6 \text{ m} \cdot 18 \text{ m}}{6 \text{ m}} = 18 \text{ m}$

33. $\dfrac{144 \text{ in}^2}{1728 \text{ in}^3} = \dfrac{144 \text{ in}^2}{144 \text{ in}^2 \cdot 12 \text{ in}} = \dfrac{1}{12 \text{ in}}$

35. $\dfrac{2060 \text{ degrees-gal}}{103 \text{ degrees}} = \dfrac{103 \text{ degrees} \cdot 20 \text{ gal}}{103 \text{ degrees}}$

                         $= 20$ gal

37. $\dfrac{7-4}{4-7} = \dfrac{3}{-3} = -1$

      Opposite numerator and denominator simplify to -1.

39. $\dfrac{-3-4}{4-(-3)} = \dfrac{-7}{7} = -1$

41. $-(a-b) = -a+b$

43. $-(-a+b) = a-b$

45. $-(b+a) = -b-a = -a-b$

47. $(n-m)+(n+m) = n+n-m+m = 2n \neq 0$;
     $-1(n-m) = -n+m$;
             $\neq n+m$
     Not opposites.

49. $(x-2)+(2-x) = x-x-2+2 = 0$;
     $-1(x-2) = -x+2 = 2-x$;
     Opposites.

51. The numerator and denominator must be equal for the equation to equal 1;
     $x+2$; $x \neq -2$

53. $-x+2$; $x \neq 2$

55. $b-a$; $a \neq b$

57. $3-x$; $x \neq 3$

59. $\dfrac{1}{4} = 4^{-1}$

61. $\dfrac{1}{a} = a^{-1}$

63. True; only factors may be simplified.

65. $\dfrac{a}{-a}, \dfrac{-a}{a}, -\dfrac{a}{a}$

67. To simplify, fractions need common factors in numerator and denominator.

Exercises 9.3

1. a. $\dfrac{1}{3} \cdot \dfrac{1}{4} = \dfrac{1}{12}$; $\dfrac{1}{3} \div \dfrac{1}{4} = \dfrac{1}{3} \cdot \dfrac{4}{1} = \dfrac{4}{3}$

   b. $\dfrac{1}{2} \cdot \dfrac{1}{5} = \dfrac{1}{10}$; $\dfrac{1}{2} \div \dfrac{1}{5} = \dfrac{1}{2} \cdot \dfrac{5}{1} = \dfrac{5}{2}$

3. a. $\dfrac{3}{4} \cdot \dfrac{1}{6} = \dfrac{3}{24} = \dfrac{1}{8}$; $\dfrac{3}{4} \div \dfrac{1}{6} = \dfrac{3}{4} \cdot \dfrac{6}{1} = \dfrac{18}{4} = \dfrac{9}{2}$

   b. $\dfrac{2}{3} \cdot \dfrac{1}{6} = \dfrac{2}{18} = \dfrac{1}{9}$; $\dfrac{2}{3} \div \dfrac{1}{6} = \dfrac{2}{3} \cdot \dfrac{6}{1} = \dfrac{12}{3} = 4$

5. a. $100 \div \dfrac{4}{1} = \dfrac{100}{1} \cdot \dfrac{1}{4} = \dfrac{100}{4} = 25$;

      $100 \cdot \dfrac{1}{4} = \dfrac{100}{1} \cdot \dfrac{1}{4} = \dfrac{100}{4} = 25$;

   b. $100 \div \dfrac{1}{5} = \dfrac{100}{1} \cdot \dfrac{5}{1} = \dfrac{500}{1} = 500$;

      $100 \cdot \dfrac{5}{1} = \dfrac{100}{1} \cdot \dfrac{5}{1} = \dfrac{500}{1} = 500$;

   Equal; division is the same as multiplication by the reciprocal.

7. a. $\dfrac{1}{x} \cdot \dfrac{x^2}{1} = \dfrac{1 \cdot x^2}{x \cdot 1} = \dfrac{1 \cdot x \cdot x}{1 \cdot x} = x$

   b. $\dfrac{1}{a} \div \dfrac{a^2 b^2}{1} = \dfrac{1}{a} \cdot \dfrac{1}{a^2 b^2} = \dfrac{1}{a^3 b^2}$

   c. $\dfrac{a}{b} \cdot \dfrac{b^2}{a^2} = \dfrac{ab^2}{ba^2} = \dfrac{a \cdot b \cdot b}{b \cdot a \cdot a} = \dfrac{b}{a}$

   d. $\dfrac{a}{b} \div \dfrac{a^2}{b^3} = \dfrac{a}{b} \cdot \dfrac{b \cdot b \cdot b}{a \cdot a} = \dfrac{b^2}{a}$

9. a. $\dfrac{x^2+2x+1}{x+1} \cdot \dfrac{x}{x^2+x} = \dfrac{(x+1)(x+1) \cdot x}{(x+1) \cdot x(x+1)}$
                                   $= 1$

   b. $\dfrac{x^2-4}{x+2} \cdot \dfrac{1}{x^2-x} = \dfrac{(x-2)(x+2) \cdot 1}{(x+2) \cdot x(x-1)}$
                                   $= \dfrac{x-2}{x(x-1)}$

   c. $\dfrac{x+2}{x^2-4x+4} \div \dfrac{x^2+2x}{x-2}$
       $= \dfrac{(x+2) \cdot (x-2)}{(x-2)(x-2) \cdot x(x+2)}$
       $= \dfrac{1}{x(x-2)}$

d. $\dfrac{x^2-5x}{x^2+5x} \div \dfrac{x^2-10x+25}{x}$

$= \dfrac{x(x-5)\cdot x}{x(x+5)\cdot(x-5)(x-5)}$

$= \dfrac{x}{(x+5)(x-5)}$

e. $\dfrac{x^2-6x+9}{x^2+3x} \div \dfrac{x^2-9}{x}$

$= \dfrac{(x-3)(x-3)\cdot x}{x(x+3)\cdot(x-3)(x+3)}$

$= \dfrac{x-3}{(x+3)^2}$

11. a. $\dfrac{x^2+x}{x-1}\cdot\dfrac{x^2-1}{x+1} = \dfrac{x(x+1)\cdot(x-1)(x+1)}{(x-1)(x+1)}$

$\qquad\qquad = x(x+1)$

b. $\dfrac{x-3}{x^2+6x+9}\cdot\dfrac{x+3}{x^2-9}$

$= \dfrac{(x-3)(x+3)}{(x+3)(x+3)(x-3)(x+3)}$

$= \dfrac{1}{(x+3)^2}$

c. $\dfrac{4-8x}{x+1} \div \dfrac{1-2x}{x^2-1}$

$= \dfrac{4(1-2x)\cdot(x-1)(x+1)}{(x+1)(1-2x)}$

$= 4(x-1)$

d. $\dfrac{x-x^2}{x+1}\cdot\dfrac{x-1}{1-x} = \dfrac{-1\cdot x(x-1)(x-1)}{(x+1)\cdot-1(x-1)}$

$\qquad\qquad = \dfrac{x(x-1)}{x+1}$

e. $\dfrac{x^2-x}{x^2-3x+2} \div \dfrac{1-x^2}{x^2-2x+1}$

$= \dfrac{x(x-1)\cdot(x-1)(x-1)}{(x-2)(x-1)\cdot-1(x-1)(x+1)}$

$= \dfrac{x(x-1)}{(x-2)(x+1)}$

13. a. $\dfrac{x}{1-x}\cdot\dfrac{x-1}{x^2} = \dfrac{x(x-1)}{-1(x-1)\cdot x^2} = -\dfrac{1}{x}$

b. $\dfrac{c^2-d^2}{d}\cdot\dfrac{cd}{d-c} = \dfrac{(c-d)(c+d)\cdot c\cdot d}{d\cdot-1(c-d)}$

$\qquad\qquad = -c(c+d)$

c. $\dfrac{x^2}{4-x} \div \dfrac{4x}{x-4} = \dfrac{x^2(x-4)}{-1(x-4)\cdot 4x} = -\dfrac{x}{4}$

d. $\dfrac{a^2-2ab+b^2}{a+b} \div \dfrac{b-a}{a+b}$

$= \dfrac{(a-b)(a-b)\cdot(a+b)}{(a+b)\cdot-1(a-b)}$

$= -(a-b)$

15. a. $\dfrac{1}{a}\cdot\dfrac{1}{b} = \dfrac{1}{ab}$

b. $a \div \dfrac{1}{a} = \dfrac{a}{1}\cdot\dfrac{a}{1} = a^2$

c. $\dfrac{1}{b}\cdot a = \dfrac{1}{b}\cdot\dfrac{a}{1} = \dfrac{a}{b}$

d. $\dfrac{1}{b} \div \dfrac{1}{a} = \dfrac{1}{b}\cdot\dfrac{a}{1} = \dfrac{a}{b}$

e. $\dfrac{a}{b}\cdot\dfrac{1}{b} = \dfrac{a}{b^2}$

f. $\dfrac{a}{b} \div \dfrac{1}{a} = \dfrac{a}{b}\cdot\dfrac{a}{1} = \dfrac{a^2}{b}$

17. $\dfrac{\text{miles}}{\dfrac{\text{miles}}{\text{hour}}} = \text{miles}\cdot\dfrac{\text{hour}}{\text{miles}} = \text{hour}$

19. $\dfrac{93{,}000{,}000\text{ mi}}{186{,}000\text{ mi per sec}} = \dfrac{93{,}000{,}000\text{ mi}}{\dfrac{186{,}000\text{ mi}}{\text{sec}}}$

$\qquad = 93{,}000{,}000\text{ mi}\cdot\dfrac{\text{sec}}{186{,}000\text{ mi}}$

$\qquad = 500\text{ sec}$

21. $\dfrac{300\text{ mi per hr}}{100\text{ gal per hr}} = \dfrac{\dfrac{300\text{ mi}}{\text{hr}}}{\dfrac{100\text{ gal}}{\text{hr}}}$

$\qquad = \dfrac{300\text{ mi}}{\text{hr}}\cdot\dfrac{\text{hr}}{100\text{ gal}}$

$\qquad = 3\text{ mpg}$

23. $\dfrac{12 \text{ cookies per dozen}}{\$2.98 \text{ per dozen}}$

$= \dfrac{\dfrac{12 \text{ cookies}}{\text{dozen}}}{\dfrac{\$2.98}{\text{dozen}}}$

$= \dfrac{12 \text{ cookies}}{\text{dozen}} \cdot \dfrac{\text{dozen}}{\$2.98}$

$\approx 4 \text{ cookies/dollar}$

25. $\dfrac{85 \text{ words per minute}}{300 \text{ words per page}} = \dfrac{\dfrac{85 \text{ words}}{\text{minute}}}{\dfrac{300 \text{ words}}{\text{page}}}$

$= \dfrac{85 \text{ words}}{\text{minute}} \cdot \dfrac{\text{page}}{300 \text{ words}}$

$\approx 0.28 \text{ page/min}$

27. $\dfrac{40 \text{ moles}}{12 \text{ moles per L}} = 40 \text{ moles} \cdot \dfrac{\text{L}}{12 \text{ moles}}$

$= 3\dfrac{1}{3} \text{ L}$

29. Answers will vary.

31. Since $I = \dfrac{V}{R}$,

$I = \dfrac{\dfrac{t^2 - 4}{2t^2 - 3t - 2}}{\dfrac{t + 2}{t^2}}$

$= \dfrac{t^2 - 4}{2t^2 - 3t - 2} \cdot \dfrac{t^2}{t + 2}$

$= \dfrac{(t - 2)(t + 2) \cdot t^2}{(2t + 1)(t - 2)(t + 2)}$

$= \dfrac{t^2}{2t + 1}$

33. We could multiply numerators and denominators and then simplify.

Mid-Chapter 9 Test

1. $x = -2, \ x = 1$

2.

Input	Output	Budget
0	undefined	4800
10	480	4800
20	240	4800
100	48	4800
1000	4.80	4800

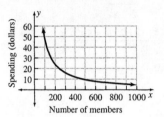

Number of members

3. $y = \dfrac{4800}{x}$, $x \neq 0$

4. y approaches $+\infty$ as x approaches -2 from the left.

5. y approaches $+\infty$ as x approaches 1 from the right.

6. y approaches $-\infty$ as x approaches 1 from the left.

7. a. $\dfrac{24ac}{28a^2} = \dfrac{4 \cdot 6 \cdot a \cdot c}{4 \cdot 7 \cdot a \cdot a} = \dfrac{6c}{7a}$

     b. No common factors.

8. a. $\dfrac{x + 2}{x^2 - 4} = \dfrac{x + 2}{(x - 2)(x + 2)} = \dfrac{1}{x - 2}$

     b. $\dfrac{x^2 - 2x + 1}{x^2 - 3x + 2} = \dfrac{(x - 1)(x - 1)}{(x - 1)(x - 2)} = \dfrac{x - 1}{x - 2}$

9. a. $\dfrac{3x}{5y} \cdot \dfrac{2x}{2x} = \dfrac{6x^2}{10xy}$

     b. $\dfrac{2a}{3b} \cdot \dfrac{4ab}{4ab} = \dfrac{8a^2b}{12ab^2}$

10. a. $\dfrac{3}{x + 5} \cdot \dfrac{5}{5} = \dfrac{15}{5(x + 5)}$

b. $\dfrac{2}{x-3} \cdot \dfrac{x+2}{x+2} = \dfrac{2(x+2)}{(x-3)(x+2)}$

11. $\dfrac{3x}{y^2} \div \dfrac{x^2}{y} = \dfrac{3x \cdot y}{y^2 \cdot x^2} = \dfrac{3}{xy}$

12. $\dfrac{2x}{y} \div \dfrac{x^2}{y} = \dfrac{2x \cdot y}{y \cdot x^2} = \dfrac{2}{x}$

13. $\dfrac{x^2-5x-6}{x+2} \cdot \dfrac{x^2-4}{2-x}$

$= \dfrac{(x-6)(x+1)(x-2)(x+2)}{(x+2) \cdot -1(x-2)}$

$= -(x-6)(x+1)$

14. $\dfrac{x^2-3x}{x^2-16} \div \dfrac{x-3}{x+4} = \dfrac{x(x-3)(x+4)}{(x-4)(x+4)(x-3)}$

$\qquad\qquad = \dfrac{x}{x-4}$

15. $\dfrac{\dfrac{2}{3x}}{\dfrac{x^2}{6}} = \dfrac{2}{3x} \cdot \dfrac{6}{x^2} = \dfrac{4}{x^3}$

16. $\dfrac{63 \text{ days}}{7 \text{ days per week}} = 63 \text{ days} \cdot \dfrac{\text{week}}{7 \text{ days}}$

$\qquad\qquad\qquad = 9 \text{ week}$

17. $\dfrac{16 \text{ stitches per sec}}{8 \text{ stitches per in}} = \dfrac{16 \text{ stitches}}{\text{sec}} \cdot \dfrac{\text{in}}{8 \text{ stitches}}$

$\qquad\qquad\qquad = 2 \text{ in/sec}$

18. $\dfrac{\dfrac{10 \text{ mL}}{100 \text{ mL}} \cdot 400 \text{ mL}}{\dfrac{50 \text{ mL}}{100 \text{ mL}}} = \dfrac{4000 \text{ mL}}{100 \text{ mL}} \cdot \dfrac{100 \text{ mL}}{50 \text{ mL}}$

$\qquad\qquad\qquad = 80 \text{ mL}$

Section 9.4

1. a. $\dfrac{11}{7} - \dfrac{4}{7} = \dfrac{11-4}{7} = \dfrac{7}{7} = 1$

b. $\dfrac{2}{3} + \dfrac{x}{3} = \dfrac{2+x}{3}$

c. $\dfrac{3}{2x} - \dfrac{5}{2x} = \dfrac{3-5}{2x} = \dfrac{-2}{2x} = -\dfrac{1}{x}$

d. $\dfrac{4}{x^2+1} - \dfrac{x^2}{x^2+1} = \dfrac{4-x^2}{x^2+1}$

e. $\dfrac{2}{x-1} - \dfrac{x+1}{x-1} = \dfrac{2-(x+1)}{x-1}$

$\qquad\qquad = \dfrac{-x+1}{x-1}$

$\qquad\qquad = \dfrac{-1(x-1)}{x-1}$

$\qquad\qquad = -1$

3. a. $12 = 2 \cdot 2 \cdot 3$

$20 = 2 \cdot 2 \cdot 5$;

LCD: $2 \cdot 2 \cdot 3 \cdot 5 = 60$

b. $x = 1 \cdot x$

$2x = 2 \cdot x$;

LCD: $2 \cdot x = 2x$

c. $y = 1 \cdot y$

$y^2 = y \cdot y$;

LCD: $y \cdot y = y^2$

d. $a = 1 \cdot a$

$b = 1 \cdot b$;

LCD: ab

e. $x - 3 = 1 \cdot (x-3)$

$x^2 - 9 = (x-3)(x+3)$;

LCD: $(x-3)(x+3)$

f. $x^2 + 5x + 6 = (x+2)(x+3)$

$x^2 - 9 = (x-3)(x+3)$;

LCD: $(x+3)(x+2)(x-3)$

5. a. $\dfrac{5}{12} + \dfrac{7}{20} = \dfrac{5 \cdot 5}{12 \cdot 5} + \dfrac{7 \cdot 3}{20 \cdot 3}$

$\qquad\qquad = \dfrac{25}{60} + \dfrac{21}{60}$

$\qquad\qquad = \dfrac{46}{60}$

$\qquad\qquad = \dfrac{23}{30}$

b. $\dfrac{2}{x} + \dfrac{5}{2x} = \dfrac{2 \cdot 2}{x \cdot 2} + \dfrac{5}{2x} = \dfrac{4}{2x} + \dfrac{5}{2x} = \dfrac{9}{2x}$

c. $\dfrac{8}{y} - \dfrac{1}{y^2} = \dfrac{8 \cdot y}{y \cdot y} - \dfrac{1}{y^2} = \dfrac{8y}{y^2} - \dfrac{1}{y^2} = \dfrac{8y-1}{y^2}$

d. $\dfrac{3}{b}+\dfrac{2}{a}+\dfrac{5}{b}-\dfrac{3}{a}$

$=\dfrac{3\cdot a}{b\cdot a}+\dfrac{2\cdot b}{a\cdot b}+\dfrac{5\cdot a}{b\cdot a}-\dfrac{3\cdot b}{a\cdot b}$

$=\dfrac{3a}{ab}+\dfrac{2b}{ab}+\dfrac{5a}{ab}-\dfrac{3b}{ab}$

$=\dfrac{8a-b}{ab}$

e. $\dfrac{4}{x-3}-\dfrac{2}{x^2-9}$

$=\dfrac{4(x+3)}{(x-3)(x+3)}-\dfrac{2}{(x-3)(x+3)}$

$=\dfrac{4x+12-2}{(x-3)(x+3)}$

$=\dfrac{4x+10}{(x-3)(x+3)}$

f. $\dfrac{4}{x^2+5x+6}-\dfrac{2}{x^2-9}$

$=\dfrac{4(x-3)-2(x+2)}{(x+2)(x+3)(x-3)}$

$=\dfrac{(4x-12)-(2x+4)}{(x+2)(x+3)(x-3)}$

$=\dfrac{2x-16}{(x-3)(x+3)(x+2)}$

7. $\dfrac{3}{2b}+\dfrac{3}{4a}=\dfrac{3\cdot 2a}{2b\cdot 2a}+\dfrac{3\cdot b}{4a\cdot b}$

$=\dfrac{6a}{4ab}+\dfrac{3b}{4ab}$

$=\dfrac{6a+3b}{4ab}$

9. $\dfrac{1}{x-1}+\dfrac{1}{x}=\dfrac{1\cdot x}{(x-1)\cdot x}+\dfrac{1(x-1)}{x(x-1)}$

$=\dfrac{x+(x-1)}{x(x-1)}$

$=\dfrac{2x-1}{x(x-1)}$

11. $\dfrac{8}{x+1}-\dfrac{3}{x}=\dfrac{8\cdot x}{(x+1)\cdot x}-\dfrac{3(x+1)}{x(x+1)}$

$=\dfrac{8x-(3x+3)}{x(x+1)}$

$=\dfrac{5x-3}{x(x+1)}$

13. $\dfrac{1}{a}-\dfrac{1}{a^2}=\dfrac{1\cdot a}{a\cdot a}-\dfrac{1}{a\cdot a}=\dfrac{a-1}{a^2}$

15. $\dfrac{2}{ab}-\dfrac{3}{2b}=\dfrac{2\cdot 2}{2\cdot ab}-\dfrac{3\cdot a}{2b\cdot a}=\dfrac{4-3a}{2ab}=\dfrac{4-3a}{2ab}$

17. a. $\dfrac{2}{x+3}+\dfrac{3}{(x+3)^2}$

$=\dfrac{2(x+3)}{(x+3)(x+3)}+\dfrac{3}{(x+3)(x+3)}$

$=\dfrac{2x+6+3}{(x+3)(x+3)}$

$=\dfrac{2x+9}{(x+3)(x+3)}$

b. $\dfrac{2}{x^2-1}-\dfrac{1}{x-1}$

$=\dfrac{2}{(x-1)(x+1)}-\dfrac{1\cdot(x+1)}{(x-1)(x+1)}$

$=\dfrac{2-(x+1)}{(x-1)(x+1)}$

$=\dfrac{-x+1}{(x-1)(x+1)}$

$=\dfrac{-1\cdot(x-1)}{(x-1)(x+1)}$

$=\dfrac{-1}{x+1}$

19. a. $\dfrac{3}{x^2-x}+\dfrac{x}{x^2-3x+2}$

$=\dfrac{3(x-2)}{x(x-1)(x-2)}+\dfrac{x\cdot x}{x(x-1)(x-2)}$

$=\dfrac{(3x-6)+x^2}{x(x-1)(x-2)}$

$=\dfrac{x^2+3x-6}{x(x-1)(x-2)}$

b.
$$\frac{1}{x^2-3x}-\frac{2}{x^2-9}$$

$$=\frac{1\cdot(x+3)}{x(x-3)(x+3)}-\frac{2\cdot x}{x(x-3)(x+3)}$$

$$=\frac{(x+3)-2x}{x(x-3)(x+3)}$$

$$=\frac{-x+3}{x(x-3)(x+3)}$$

$$=\frac{-1\cdot(x-3)}{x(x-3)(x+3)}$$

$$=\frac{x-3}{x(x+3)}$$

21. $\Delta T = \dfrac{T_0}{T}-1=\dfrac{T_0}{T}-\dfrac{T}{T}=\dfrac{T_0-T}{T}$

23. $\dfrac{1}{C}=\dfrac{1}{C_1}+\dfrac{1}{C_2}=\dfrac{1\cdot C_2}{C_1\cdot C_2}+\dfrac{1\cdot C_1}{C_2\cdot C_1}=\dfrac{C_2+C_1}{C_1 C_2}$

25. $t=\dfrac{D}{r_1}+\dfrac{D}{r_2}$

$$=\frac{D\cdot r_2}{r_1\cdot r_2}+\frac{D\cdot r_1}{r_2\cdot r_1}$$

$$=\frac{Dr_2}{r_1 r_2}+\frac{Dr_1}{r_1 r_2}$$

$$=\frac{D(r_1+r_2)}{r_1 r_2}$$

27. $F=\dfrac{L^2}{6d}+\dfrac{d}{2}=\dfrac{L^2}{6d}+\dfrac{d\cdot 3d}{2\cdot 3d}=\dfrac{L^2+3d^2}{6d}$

29. $m=\dfrac{\dfrac{1}{3}-\dfrac{4}{3}}{\dfrac{1}{3}-\dfrac{2}{3}}=\dfrac{-\dfrac{5}{3}}{-\dfrac{1}{3}}=-\dfrac{5}{3}\cdot-\dfrac{3}{1}=\dfrac{15}{3}=5$

31. $m=\dfrac{\dfrac{1}{5}-\dfrac{2}{5}}{\dfrac{-4}{5}-\dfrac{1}{5}}=\dfrac{-\dfrac{1}{5}}{-\dfrac{5}{5}}=-\dfrac{1}{5}\cdot\left(-\dfrac{5}{5}\right)=\dfrac{1}{5}$

33. $m=\dfrac{\dfrac{a}{2}-a}{b-\left(-\dfrac{b}{2}\right)}=\dfrac{-\dfrac{a}{2}}{\dfrac{3b}{2}}=\left(-\dfrac{a}{2}\right)\cdot\dfrac{2}{3b}=-\dfrac{a}{3b}$

35. $m=\dfrac{b-0}{a-c}=\dfrac{b}{a-c}$;

$m=\dfrac{\dfrac{b}{2}-0}{\dfrac{a}{2}-\dfrac{c}{2}}=\dfrac{\dfrac{b}{2}}{\dfrac{a-c}{2}}=\dfrac{b}{2}\cdot\dfrac{2}{a-c}=\dfrac{b}{a-c}$;

The slopes are equal.

37. $\dfrac{5}{\frac{1}{5}+1}=\dfrac{\frac{5}{1}\cdot 5}{\frac{6}{5}\cdot 5}=\dfrac{25}{6}$

39. $h=\dfrac{A}{\frac{1}{2}b}=\dfrac{A\cdot 2}{\frac{1}{2}b\cdot 2}=\dfrac{2A}{b}$

41. $\dfrac{x+\dfrac{x}{2}}{2-\dfrac{x}{3}}=\dfrac{\dfrac{3x}{2}\cdot 6}{\dfrac{6-x}{3}\cdot 6}=\dfrac{\dfrac{3x}{1}\cdot 3}{\dfrac{6-x}{1}\cdot 2}=\dfrac{9x}{2(6-x)}$

43. $\dfrac{\dfrac{1}{Q_H}}{\dfrac{Q_H}{Q_L}-1}=\dfrac{1\cdot Q_L}{\dfrac{Q_H-Q_L}{Q_L}\cdot Q_L}=\dfrac{Q_L}{Q_H-Q_L}$

45. a.

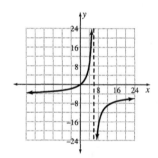

b. The graphs before and after simplifying are the same.

c. The graph is vertical near $x=6$ because the expression is undefined when $x=6$.

47. a. $\dfrac{3}{4}\cdot\dfrac{2}{5}=\dfrac{6}{20}=\dfrac{3}{10}$

b. $\dfrac{3}{4}\div\dfrac{2}{5}=\dfrac{3}{4}\cdot\dfrac{5}{2}=\dfrac{15}{8}$

c. $\dfrac{3}{4}+\dfrac{2}{5}=\dfrac{3\cdot 5}{4\cdot 5}+\dfrac{2\cdot 4}{5\cdot 4}=\dfrac{15}{20}+\dfrac{8}{20}=\dfrac{23}{20}$

49. a. $\dfrac{1}{a}+\dfrac{1}{b}=\dfrac{1\cdot b}{a\cdot b}+\dfrac{1\cdot a}{b\cdot a}=\dfrac{b}{ab}+\dfrac{a}{ab}=\dfrac{a+b}{ab}$

b. $\dfrac{1}{a}\cdot\dfrac{1}{b}=\dfrac{1}{ab}$

Exercises 9.5

1. $12x\left(\dfrac{1}{12}+\dfrac{2}{3x}\right)=\dfrac{12x}{12}+\dfrac{12x\cdot 2}{3x}$

$\qquad\qquad\quad=\dfrac{12x}{12}+\dfrac{24x}{3x}$

$\qquad\qquad\quad=x+8$

3. $4x^2\left(\dfrac{1}{2x}+\dfrac{3}{x^2}\right)=\dfrac{4x^2}{2x}+\dfrac{4x^2\cdot 3}{x^2}$

$\qquad\qquad\quad=\dfrac{4x^2}{2x}+\dfrac{12x^2}{x^2}$

$\qquad\qquad\quad=2x+12$

5. $x(x-1)\left(\dfrac{1}{x-1}+\dfrac{1}{x}\right)=\dfrac{x(x-1)}{x-1}+\dfrac{x(x-1)}{x}$

$\qquad\qquad\qquad\quad=x+(x-1)$

$\qquad\qquad\qquad\quad=x+x-1$

$\qquad\qquad\qquad\quad=2x-1$

7. $\dfrac{2}{(2)}+\dfrac{3(2)}{2}\overset{?}{=}4$

$\qquad\dfrac{2}{2}+\dfrac{6}{2}\overset{?}{=}4$

$\qquad\quad 1+3\overset{?}{=}4$

$\qquad\qquad 4=4\ \ \text{true}$

$x=2$ is a solution.

9. $\dfrac{2}{\left(-\frac{1}{2}\right)+1}+\dfrac{3}{\left(-\frac{1}{2}\right)}\overset{?}{=}-2$

$\qquad\dfrac{2}{\frac{1}{2}}-\dfrac{3}{\frac{1}{2}}\overset{?}{=}-2$

$\qquad\quad 4-6\overset{?}{=}-2$

$\qquad\qquad -2=-2\ \ \text{true}$

$x=-\dfrac{1}{2}$ is a solution.

11. $4=2\cdot 2$

$6=2\cdot 3$

$\text{LCD}=2\cdot 2\cdot 3=12$

13. $4=2\cdot 2$

5 is prime

x is prime

$\text{LCD}=2\cdot 2\cdot 5\cdot x=20x$

15. $8=2\cdot 2\cdot 2$

2 is prime

x is prime

$\text{LCD}=2\cdot 2\cdot 2\cdot x=8x$

17. x is prime

$3x=3\cdot x$

3 is prime

$\text{LCD}=3\cdot x=3x$

19. $\text{LCD}=x$

21. $(x-1)$ is prime

$(x+3)$ is prime

$\text{LCD}=(x-1)(x+3)$

23. $x^2=x\cdot x$

x is prime

$\text{LCD}=x\cdot x=x^2$

25. $\text{LCD}=x-3$

27. $\dfrac{x}{4}+\dfrac{x}{6}=28$

$12\left(\dfrac{x}{4}+\dfrac{x}{6}\right)=12(28)$

$\qquad 3x+2x=336$

$\qquad\qquad 5x=336$

$\qquad\qquad x=\dfrac{336}{5}\ \text{ or }\ 67.2$

29. $\dfrac{3}{4}+\dfrac{1}{5}=\dfrac{1}{x};\ \ x\neq 0$

$20x\left(\dfrac{3}{4}+\dfrac{1}{5}\right)=20x\left(\dfrac{1}{x}\right)$

$\qquad 15x+4x=20$

$\qquad\qquad 19x=20$

$\qquad\qquad x=\dfrac{20}{19}$

31. $\dfrac{1}{8}+\dfrac{1}{x}=\dfrac{1}{2};\quad x\neq 0$

$8x\left(\dfrac{1}{8}+\dfrac{1}{x}\right)=8x\left(\dfrac{1}{2}\right)$

$x+8=4x$

$8=3x$

$\dfrac{8}{3}=x\quad\text{or}\quad x=\dfrac{8}{3}$

33. $\dfrac{1}{x}=\dfrac{1}{3x}+\dfrac{1}{3};\quad x\neq 0$

$3x\left(\dfrac{1}{x}\right)=3x\left(\dfrac{1}{3x}+\dfrac{1}{3}\right)$

$3=1+x$

$2=x\quad\text{or}\quad x=2$

35. $\dfrac{3}{x}-\dfrac{2}{x}=\dfrac{4}{x};\quad x\neq 0$

$x\left(\dfrac{3}{x}-\dfrac{2}{x}\right)=x\left(\dfrac{4}{x}\right)$

$3-2=4$

$1=4\quad\text{false}$

The equation has no solution.

37. $\dfrac{1}{x-1}=\dfrac{2}{x+3};\quad x\neq 1, x\neq -3$

$(x-1)(x+3)\left(\dfrac{1}{x-1}\right)=(x-1)(x+3)\left(\dfrac{2}{x+3}\right)$

$x+3=2(x-1)$

$x+3=2x-2$

$5=x\quad\text{or}\quad x=5$

39. $\dfrac{2}{x^2}-\dfrac{3}{x}+1=0;\quad x\neq 0$

$x^2\left(\dfrac{2}{x^2}-\dfrac{3}{x}+1\right)=x^2(0)$

$2-3x+x^2=0$

$(x-1)(x-2)=0$

$x-1=0\quad\text{or}\quad x-2=0$

$x=1\quad\text{or}\quad x=2$

41. $\dfrac{1}{x-3}-3=\dfrac{4-x}{x-3};\quad x\neq 3$

$(x-3)\left(\dfrac{1}{x-3}-3\right)=(x-3)\left(\dfrac{4-x}{x-3}\right)$

$1-3(x-3)=4-x$

$1-3x+9=4-x$

$6=2x$

$3=x\quad\text{excluded}$

The equation has no solution.

43. $5x+\dfrac{13}{2}=\dfrac{3}{2x};\quad x\neq 0$

$2x\left(5x+\dfrac{13}{2}\right)=2x\left(\dfrac{3}{2x}\right)$

$10x^2+13x=3$

$10x^2+13x-3=0$

$(5x-1)(2x+3)=0$

$5x-1=0\quad\text{or}\quad 2x+3=0$

$5x=1\quad\text{or}\quad 2x=-3$

$x=\dfrac{1}{5}\quad\text{or}\quad x=-\dfrac{3}{2}$

45. $\dfrac{1}{x}+3=\dfrac{5}{x(x+1)};\quad x\neq 0, x\neq -1$

$x(x+1)\left(\dfrac{1}{x}+3\right)=x(x+1)\left(\dfrac{5}{x(x+1)}\right)$

$x+1+3x(x+1)=5$

$x+1+3x^2+3x=5$

$3x^2+4x-4=0$

$(3x-2)(x+2)=0$

$3x-2=0\quad\text{or}\quad x+2=0$

$3x=2\quad\text{or}\quad x=-2$

$x=\dfrac{2}{3}\quad\text{or}\quad x=-2$

47.
$$x = 6 - \frac{6}{x-1}; \quad x \neq 1$$
$$(x-1)(x) = (x-1)\left(6 - \frac{6}{x-1}\right)$$
$$x^2 - x = 6(x-1) - 6$$
$$x^2 - x = 6x - 6 - 6$$
$$x^2 - 7x + 12 = 0$$
$$(x-4)(x-3) = 0$$
$$x - 4 = 0 \text{ or } x - 3 = 0$$
$$x = 4 \text{ or } x = 3$$

49.
$$\frac{x-1}{x-2} = \frac{1}{x-2}; \quad x \neq 2$$
$$(x-2)\left(\frac{x-1}{x-2}\right) = (x-2)\left(\frac{1}{x-2}\right)$$
$$x - 1 = 1$$
$$\cancel{x=2} \text{ excluded}$$
The equation has no solution.

51.
$$\frac{10}{x-5} = x + \frac{2x}{x-5}; \quad x \neq 5$$
$$(x-5)\left(\frac{10}{x-5}\right) = (x-5)\left(x + \frac{2x}{x-5}\right)$$
$$10 = x(x-5) + 2x$$
$$10 = x^2 - 5x + 2x$$
$$0 = x^2 - 3x - 10$$
$$0 = (x-5)(x+2)$$
$$x - 5 = 0 \text{ or } x + 2 = 0$$
$$\cancel{x=5} \text{ or } x = -2$$
The only solution is $x = -2$.

53.
$$2x + \frac{2x}{x+1} = \frac{5}{x+1}; \quad x \neq -1$$
$$(x+1)\left(2x + \frac{2x}{x+1}\right) = (x+1)\left(\frac{5}{x+1}\right)$$
$$2x(x+1) + 2x = 5$$
$$2x^2 + 2x + 2x - 5 = 0$$
$$2x^2 + 4x - 5 = 0$$
Use the quadratic equation to solve.

$$x = \frac{-4 \pm \sqrt{4^2 - 4(2)(-5)}}{2(2)}$$
$$= \frac{-4 \pm \sqrt{16 + 100}}{4}$$
$$= \frac{-4 \pm \sqrt{116}}{4}$$
$$= \frac{-4 \pm 2\sqrt{29}}{4}$$
$$= \frac{-2 \pm \sqrt{29}}{2}$$

55.
$$\frac{1}{15} + \frac{1}{18} = \frac{1}{d}$$
$$90d\left(\frac{1}{15} + \frac{1}{18}\right) = 90d\left(\frac{1}{d}\right)$$
$$6d + 5d = 90$$
$$11d = 90$$
$$d = \frac{90}{11}$$

57.
$$\frac{1}{8} + \frac{1}{20} = \frac{1}{x}$$
$$40x\left(\frac{1}{8} + \frac{1}{20}\right) = 40x\left(\frac{1}{x}\right)$$
$$5x + 2x = 40$$
$$7x = 40$$
$$x = \frac{40}{7}$$

59. Let d be the number of days it takes them to work together. Adding their work rates, we get:
$$\frac{1}{14} + \frac{1}{12} = \frac{1}{d}$$
$$84d\left(\frac{1}{14} + \frac{1}{12}\right) = 84d\left(\frac{1}{d}\right)$$
$$6d + 7d = 84$$
$$13d = 84$$
$$d = \frac{84}{13} \approx 6.5$$
It will take about 6.5 days for the two farmers to harvest the crop together.

61. Let t be the number of hours needed for the two fans working together to vent the house. Adding the rates we get:

$$\frac{1}{4}+\frac{1}{5}=\frac{1}{t}$$

$$20t\left(\frac{1}{4}+\frac{1}{5}\right)=20t\left(\frac{1}{t}\right)$$

$$5t+4t=20$$

$$9t=20$$

$$t=\frac{20}{9}\approx 2.2$$

It will take the two fans about 2.2 hours to vent the house if both fans work together. Since this is less than 3 hours, it satisfies the building code.

63. Let t be the number of hours it takes the second fan to vent the house if it worked by itself. Adding the rates we get:

$$\frac{1}{5}+\frac{1}{t}=\frac{1}{3}$$

$$\frac{1}{t}=\frac{1}{3}-\frac{1}{5}$$

$$15t\left(\frac{1}{t}\right)=15t\left(\frac{1}{3}-\frac{1}{5}\right)$$

$$15=5t-3t$$

$$15=2t$$

$$7.5=t$$

The second fan must be able to vent the house by itself in 7.5 hours.

65. Let m be the number of minutes needed to empty the theater if both doors are used. Adding the rates we get:

$$\frac{1}{9}+\frac{1}{6}=\frac{1}{m}$$

$$18m\left(\frac{1}{9}+\frac{1}{6}\right)=18m\left(\frac{1}{m}\right)$$

$$2m+3m=18$$

$$5m=18$$

$$m=\frac{18}{5}=3.6$$

It will take 3.6 minutes to empty the theater if both doors are used.

67.
$$\frac{1}{a}+\frac{1}{b}=\frac{1}{c}$$

$$abc\left(\frac{1}{a}+\frac{1}{b}\right)=abc\left(\frac{1}{c}\right)$$

$$bc+ac=ab$$

$$ac=ab-bc$$

$$ac=b(a-c)$$

$$\frac{ac}{a-c}=b$$

69.
$$\frac{1}{a}+\frac{1}{b}=\frac{1}{c}$$

$$abc\left(\frac{1}{a}+\frac{1}{b}\right)=abc\left(\frac{1}{c}\right)$$

$$bc+ac=ab$$

$$c(a+b)=ab$$

$$c=\frac{ab}{a+b}$$

71. $[\,(\,]\,[\,1\,]\,[\div]\,[\,1\,]\,[\,5\,]\,[+]\,[\,1\,]\,[\div]\,[\,1\,]\,[\,8\,]$

$[\,)\,]\,[\,x^{-1}\,]$ [ENTER]

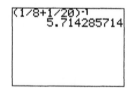

73. $[\,(\,]\,[\,1\,]\,[\div]\,[\,8\,]\,[+]\,[\,1\,]\,[\div]\,[\,2\,]\,[\,0\,]$

$[\,)\,]\,[\,x^{-1}\,]$ [ENTER]

```
(1/8+1/20)⁻¹
       5.714285714
```

75. In both cases, multiplying by the LCD eliminates the denominators.

Chapter 9 Review Exercises

1. The expression is undefined when $x = -3$ since that input value makes the denominator equal 0.

3. Let y be the number of years and x be the number of cubic feet of gas used each day. Then,
$$y = \frac{5000 \text{ trillion}}{365x}$$

5. a. As x approaches -1 from the right, the graph goes off to $-\infty$.

   b. As x approaches -1 from the left, the graph goes off to $-\infty$.

7. $\dfrac{2xy}{x^2} = \dfrac{2y \cdot x}{x \cdot x} = \dfrac{2y}{x}$

9. $\dfrac{a^2 - b^2}{a+b} = \dfrac{(a-b)(a+b)}{a+b} = a - b$

11. $\dfrac{ab}{a^2 + b^2}$
    This expression does not reduce because there are no common factors.

13. $\dfrac{1-a}{a-1} = \dfrac{-(a-1)}{a-1} = -1$; $a \neq 1$

15. $-(x-3) = 3 - x$

17. $a - b$

19. $-(4-x) = x - 4$

21. The fractions $\dfrac{16}{9}$ and $\dfrac{\sqrt{16}}{\sqrt{9}}$ are not equal since they do not simplify to the same fraction.
$$\frac{\sqrt{16}}{\sqrt{9}} = \frac{4}{3}$$

23. a. $300 \div \dfrac{3}{1} = \dfrac{300}{3} = 100$
    $$300 \cdot \frac{1}{3} = 100$$

    b. They both equal 100.

c. Division by a number a is the same as multiplying by the number $\dfrac{1}{a}$.

25. The word 'each' implies division.
$$\frac{4}{1/3} = 4 \cdot \frac{3}{1} = 12$$
There would be 12 servings in 4 large pizzas.

27. The word 'half' can imply division by 2 or multiplication by $\frac{1}{2}$.
$$\frac{1}{2} \cdot \frac{3}{4} = \frac{3}{8}$$
Jim ran $\dfrac{3}{8}$ of a mile.

29. The word 'remains' indicates subtraction.
$$\frac{1}{2} + \frac{1}{3} = \frac{3}{6} + \frac{2}{6} = \frac{5}{6}$$
$$1 - \frac{5}{6} = \frac{6}{6} - \frac{5}{6} = \frac{1}{6}$$
The shelter needs to find $\dfrac{1}{6}$ of its funding through private donations.

31. $\dfrac{1-x}{x+1} \div \dfrac{x^2-1}{x^2} = \dfrac{1-x}{x+1} \cdot \dfrac{x^2}{x^2-1}$
$$= \frac{-(x-1) \cdot x^2}{(x+1)(x+1)(x-1)}$$
$$= \frac{-\cancel{(x-1)} \cdot x^2}{(x+1)(x+1)\cancel{(x-1)}}$$
$$= -\frac{x^2}{(x+1)^2}$$

33. $\dfrac{n-2}{n(n-1)} \cdot \dfrac{(n+1)n(n-1)}{n-2}$
$$= \frac{n(n-2)(n+1)(n-1)}{n(n-1)(n-2)}$$
$$= \frac{\cancel{n}\cancel{(n-2)}(n+1)\cancel{(n-1)}}{\cancel{n}\cancel{(n-1)}\cancel{(n-2)}}$$
$$= n + 1$$

35. $\dfrac{9-3x}{x+3}\cdot\dfrac{x}{x^2-6x+9}$

$=\dfrac{-3(x-3)\cdot x}{(x+3)(x-3)(x-3)}$

$=\dfrac{-3\cancel{(x-3)}\cdot x}{(x+3)(x-3)\cancel{(x-3)}}$

$=-\dfrac{3x}{(x+3)(x-3)}$

37. $\dfrac{x}{x+2}-\dfrac{2}{x^2-4}=\dfrac{x}{x+2}-\dfrac{2}{(x-2)(x+2)}$

$=\dfrac{x(x-2)-2}{(x+2)(x-2)}$

$=\dfrac{x^2-2x-2}{(x+2)(x-2)}$

39. $\dfrac{a}{a+b}-\dfrac{b}{a-b}=\dfrac{a(a-b)-b(a+b)}{(a+b)(a-b)}$

$=\dfrac{a^2-ab-ab-b^2}{(a+b)(a-b)}$

$=\dfrac{a^2-2ab-b^2}{(a+b)(a-b)}$

41. $1-\dfrac{x^2}{2}+\dfrac{x^4}{24}-\dfrac{x^6}{720}$

$=\dfrac{720}{720}-\dfrac{360x^2}{720}+\dfrac{30x^4}{720}-\dfrac{x^6}{720}$

$=\dfrac{720-360x^2+30x^4-x^6}{720}$

43. $\dfrac{1}{a}+\dfrac{2a}{3}\cdot\dfrac{6}{a}\div\dfrac{1}{3}-\dfrac{a}{3}$

$=\dfrac{1}{a}+\dfrac{4}{1}\div\dfrac{1}{3}-\dfrac{a}{3}$

$=\dfrac{1}{a}+\dfrac{12}{1}-\dfrac{a}{3}$

$=\dfrac{3+36a-a^2}{3a}$

45. $\dfrac{4\text{ buttons per card}}{12\text{ buttons per shirt}}=\dfrac{4\text{ buttons}}{1\text{ card}}\div\dfrac{12\text{ buttons}}{1\text{ shirt}}$

$=\dfrac{4\text{ buttons}}{1\text{ card}}\cdot\dfrac{1\text{ shirt}}{12\text{ buttons}}$

$=\dfrac{1}{3}\text{ shirt/card}$

47. gr $\dfrac{1}{2}\cdot\dfrac{1\text{ tab}}{\text{gr }\frac{1}{6}}=\dfrac{\frac{1}{2}\text{ tab}}{\frac{1}{6}}=\dfrac{1}{2}\cdot\dfrac{6}{1}\text{ tab}=3\text{ tab}$

49. gr $\dfrac{1}{150}\cdot\dfrac{1\text{ mL}}{\text{gr }\frac{1}{750}}=\dfrac{\frac{1}{150}\text{ mL}}{\frac{1}{750}}$

$=\dfrac{1}{150}\cdot\dfrac{750}{1}\text{ mL}$

$=3\text{ mL}$

51. $\dfrac{d}{\frac{1}{2}g}=\dfrac{d}{1}\cdot\dfrac{2}{1g}=\dfrac{2d}{g}$

53. $(x+2)(x-2)\left(\dfrac{2}{x-2}+\dfrac{1}{x+2}\right)$

$=2(x+2)+1(x-2)$

$=2x+4+x-2$

$=3x+2$

55. $\dfrac{1}{5}+\dfrac{1}{x}=\dfrac{8}{5x};\ \ x\neq0$

$5x\left(\dfrac{1}{5}+\dfrac{1}{x}\right)=5x\left(\dfrac{8}{5x}\right)$

$x+5=8$

$x=3$

57. $\dfrac{1}{2x}=\dfrac{2}{x}-\dfrac{x}{24};\ \ x\neq0$

$24x\left(\dfrac{1}{2x}\right)=24x\left(\dfrac{2}{x}-\dfrac{x}{24}\right)$

$12=48-x^2$

$x^2=36$

$x=\pm6$

59. $\dfrac{1}{m_1}+\dfrac{1}{m_2}=\dfrac{m_2}{m_1m_2}+\dfrac{m_1}{m_1m_2}=\dfrac{m_2+m_1}{m_1m_2}$

61. hot rate + cold rate = combined rate

$$\frac{1\text{ fill}}{3\text{ min}}+\frac{1\text{ fill}}{2\text{ min}}=\frac{1\text{ fill}}{t\text{ min}}$$

$$\frac{1}{3}+\frac{1}{2}=\frac{1}{t}$$

$$\frac{2}{6}+\frac{3}{6}=\frac{1}{t}$$

$$\frac{5}{6}=\frac{1}{t}$$

$$t=\frac{6}{5}$$

It will take both lines 1.2 minutes to fill the washing machine.

63. Floor sizes requiring less than a full can of paint will lie between the curve and the axes.

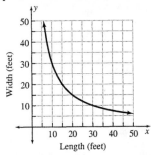

Chapter 9 Test

1. The expression is undefined for values of x that make the denominator 0.

$$x-4=0$$

$$x=4$$

2. Since $4-x$ is the opposite of $x-4$, the quotient is -1.

3. For $x=-5$,

$$y=\frac{4}{-5-2}=\frac{4}{-7}=-\frac{4}{7}$$

For $x=-4$,

$$y=\frac{4}{-4-2}=\frac{4}{-6}=-\frac{2}{3}$$

For $x=-3$,

$$y=\frac{4}{-3-2}=\frac{4}{-5}=-\frac{4}{5}$$

For $x=-2$,

$$y=\frac{4}{-2-2}=\frac{4}{-4}=-1$$

For $x=-1$,

$$y=\frac{4}{-1-2}=\frac{4}{-3}=-\frac{4}{3}$$

For $x=0$,

$$y=\frac{4}{0-2}=\frac{4}{-2}=-2$$

For $x=1$,

$$y=\frac{4}{1-2}=\frac{4}{-1}=-4$$

For $x=2$,

$$y=\frac{4}{2-2}=\frac{4}{0}=\text{undefined}$$

For $x=3$,

$$y=\frac{4}{3-2}=\frac{4}{1}=4$$

For $x=4$,

$$y=\frac{4}{4-2}=\frac{4}{2}=2$$

For $x=5$,

$$y=\frac{4}{5-2}=\frac{4}{3}$$

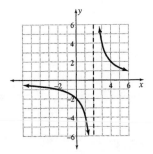

4. a. $\dfrac{6\div3}{9\div3}=\dfrac{2}{3}=\dfrac{6}{9}$

b. $\dfrac{6-3}{9-3}=\dfrac{3}{6}=\dfrac{1}{2}\neq\dfrac{6}{9}$

c. $\dfrac{6+3}{9+3}=\dfrac{9}{12}=\dfrac{3}{4}\neq\dfrac{6}{9}$

d. $\dfrac{6\cdot3}{9\cdot3}=\dfrac{18}{27}=\dfrac{2}{3}=\dfrac{6}{9}$

a and d are equivalent to $\frac{6}{9}$ because of the simplification property of fractions.

5. $\dfrac{b^2}{3ab}=\dfrac{b\cancel{b}}{3a\cancel{b}}=\dfrac{b}{3a}$

6. $\dfrac{a-b}{ab}$ cannot be simplified because there are no common factors.

7. $\dfrac{a^2-b^2}{a-b}=\dfrac{(a-b)(a+b)}{a-b}$

$=\dfrac{\cancel{(a-b)}(a+b)}{\cancel{a-b}}=\dfrac{a+b}{1}=a+b$

8. $\dfrac{xy}{xy+y}=\dfrac{xy}{y(x+1)}=\dfrac{x\cancel{y}}{\cancel{y}(x+1)}=\dfrac{x}{x+1}$

9. $\dfrac{1-x}{x+1}\cdot\dfrac{x^2-1}{x^2}=\dfrac{1-x}{x+1}\cdot\dfrac{(x+1)(x-1)}{x^2}$

$=\dfrac{1-x}{\cancel{x+1}}\cdot\dfrac{\cancel{(x+1)}(x-1)}{x^2}=\dfrac{(1-x)(x-1)}{x^2}$

or $\dfrac{-1(x-1)^2}{x^2}$

10. $\dfrac{x-1}{1-x^2}\div\dfrac{x^2-2x+1}{1+x}=\dfrac{x-1}{1-x^2}\cdot\dfrac{1+x}{x^2-2x+1}$

$=\dfrac{x-1}{(1-x)(1+x)}\cdot\dfrac{1+x}{(x-1)(x-1)}$

$=\dfrac{\cancel{x-1}}{(1-x)\cancel{(1+x)}}\cdot\dfrac{\cancel{1+x}}{\cancel{(x-1)}(x-1)}$

$=\dfrac{1}{(1-x)(x-1)}$ or $\dfrac{-1}{(x-1)(x-1)}=\dfrac{-1}{(x-1)^2}$

11. $\dfrac{n+1}{n}\div\dfrac{n(n-1)}{n^2}=\dfrac{n+1}{n}\cdot\dfrac{n^2}{n(n-1)}$

$=\dfrac{(n+1)n^2}{n^2(n-1)}=\dfrac{(n+1)\cancel{n^2}}{\cancel{n^2}(n-1)}=\dfrac{n+1}{n-1}$

12. $\dfrac{x^2-2x}{x^2-4}\cdot\dfrac{x-2}{x}=\dfrac{x(x-2)}{(x-2)(x+2)}\cdot\dfrac{x-2}{x}$

$=\dfrac{\cancel{x}\cancel{(x-2)}}{\cancel{(x-2)}(x+2)}\cdot\dfrac{x-2}{\cancel{x}}=\dfrac{x-2}{x+2}$

13. $(x+1)(x-1)\left(\dfrac{1}{x+1}+\dfrac{2}{x-1}\right)$

$=(x+1)(x-1)\cdot\dfrac{1}{x+1}$

$\quad+(x+1)(x-1)\cdot\dfrac{2}{x-1}$

$=x-1+2(x+1)$

$=x-1+2x+2=3x+1$

14. $\dfrac{4\text{ yards}}{\text{shirt}}\cdot\dfrac{\$8.98}{\text{yard}}=\dfrac{4\text{ \cancel{yards}}}{\text{shirt}}\cdot\dfrac{\$8.98}{\cancel{\text{yard}}}$

$=\$35.92/\text{shirt}$

15. $\dfrac{\dfrac{\$2.50}{1\text{ gallon}}}{\dfrac{25\text{ miles}}{1\text{ gallon}}}=\dfrac{\$2.50}{1\text{ gallon}}\div\dfrac{25\text{ miles}}{1\text{ gallon}}$

$\dfrac{\$2.50}{1\text{ gallon}}\cdot\dfrac{1\text{ gallon}}{25\text{ miles}}=\dfrac{\$2.50}{1\text{ \cancel{gallon}}}\cdot\dfrac{1\text{ \cancel{gallon}}}{25\text{ miles}}$

$=\$0.10/\text{mile}$

16. $3\text{ mg}\cdot\dfrac{\text{gr }1}{60\text{ mg}}\cdot\dfrac{1\text{ tab}}{\text{gr }\frac{1}{120}}$

$=\dfrac{3\text{ \cancel{mg}}}{1}\cdot\dfrac{\text{\cancel{gr} }1}{60\text{ \cancel{mg}}}\cdot\dfrac{1\text{ tab}}{\text{\cancel{gr} }\frac{1}{120}}=6\text{ tab}$

17. Let $x=$ rate of new door.

$\dfrac{1}{16}+\dfrac{1}{x}=\dfrac{1}{5}$

$80x\left(\dfrac{1}{16}+\dfrac{1}{x}\right)=80x\cdot\dfrac{1}{5}$

$5x+80=16x$

$80=11x$

$7.3\approx x$

The new door can empty the meeting room in about 7.3 minutes.

18. $\dfrac{1}{p}+\dfrac{1}{q}=\dfrac{1}{p}\cdot\dfrac{q}{q}+\dfrac{1}{q}\cdot\dfrac{p}{p}=\dfrac{q}{pq}+\dfrac{p}{pq}$

$=\dfrac{q+p}{pq}$

19. $\dfrac{A}{\frac{1}{2}b}=A\div\frac{1}{2}b=A\div\frac{b}{2}=A\cdot\frac{2}{b}=\frac{2A}{b}$

20. $\dfrac{2}{9}+\dfrac{7}{15}=\dfrac{2}{9}\cdot\dfrac{5}{5}+\dfrac{7}{15}\cdot\dfrac{3}{3}=\dfrac{10}{45}+\dfrac{21}{45}=\dfrac{31}{45}$

21. $\dfrac{2}{ab^2}-\dfrac{a}{2b}=\dfrac{2}{ab^2}\cdot\dfrac{2}{2}-\dfrac{a}{2b}\cdot\dfrac{ab}{ab}$

$=\dfrac{4}{2ab^2}-\dfrac{a^2b}{2ab^2}=\dfrac{4-a^2b}{2ab^2}$

22.
$$\frac{x}{x+2}+\frac{2}{x^2-4}=\frac{x}{x+2}+\frac{2}{(x+2)(x-2)}$$

$$\frac{x}{x+2}\cdot\frac{x-2}{x-2}+\frac{2}{(x+2)(x-2)}$$

$$=\frac{x^2-2x}{(x+2)(x-2)}+\frac{2}{(x+2)(x-2)}$$

$$=\frac{x^2-2x+2}{(x+2)(x-2)}$$

23.
$$\frac{2}{3}+\frac{3}{x}-\frac{4}{x^2}=\frac{2}{3}\cdot\frac{x^2}{x^2}+\frac{3}{x}\cdot\frac{3x}{3x}-\frac{4}{x^2}\cdot\frac{3}{3}$$

$$=\frac{2x^2}{3x^2}+\frac{9x}{3x^2}-\frac{12}{3x^2}=\frac{2x^2+9x-12}{3x^2}$$

24.
$$\frac{a}{2}+\frac{2}{3}\cdot\frac{a}{2}-\frac{3a}{2}\div\frac{9}{2}+\frac{1}{4}$$

$$=\frac{a}{2}+\frac{a}{3}-\frac{3a}{2}\div\frac{9}{2}+\frac{1}{4}$$

$$=\frac{a}{2}+\frac{a}{3}-\frac{3a}{2}\cdot\frac{2}{9}+\frac{1}{4}$$

$$=\frac{a}{2}+\frac{a}{3}-\frac{a}{3}+\frac{1}{4}=\frac{a}{2}+\frac{1}{4}=\frac{a}{2}\cdot\frac{2}{2}+\frac{1}{4}$$

$$=\frac{2a}{4}+\frac{1}{4}=\frac{2a+1}{4}$$

25. The input that must be excluded is $x=0$.

$$\frac{1}{3}+\frac{1}{6}=\frac{1}{x}$$

$$6x\left(\frac{1}{3}+\frac{1}{6}\right)=6x\cdot\frac{1}{x}$$

$$2x+x=6$$

$$3x=6$$

$$x=2$$

26. The input that must be excluded is $x=0$.

$$\frac{1}{x}=\frac{1}{2x}+\frac{1}{2}$$

$$2x\cdot\frac{1}{x}=2x\left(\frac{1}{2x}+\frac{1}{2}\right)$$

$$2=1+x$$

$$1=x$$

27. The inputs that must be excluded are $x=0$ and $x=3$.

$$\frac{3}{x}=\frac{2}{x-3}$$

$$x(x-3)\cdot\frac{3}{x}=x(x-3)\cdot\frac{2}{x-3}$$

$$3(x-3)=2x$$

$$3x-9=2x$$

$$x-9=0$$

$$x=9$$

28. The input that must excluded is $x=0$.

$$\frac{x-2}{4}+\frac{1}{x}=\frac{19}{4x}$$

$$4x\left(\frac{x-2}{4}+\frac{1}{x}\right)=4x\cdot\frac{19}{4x}$$

$$x(x-2)+4=19$$

$$x^2-2x+4=19$$

$$x^2-2x-15=0$$

$$(x-5)(x+3)=0$$

$$x-5=0 \quad\text{and}\quad x+3=0$$

$$x=5 \quad\text{and}\quad x=-3$$

29. The input that must excluded is $x=2$.

$$x+\frac{10}{x-2}=\frac{5x}{x-2}$$

$$(x-2)\left(x+\frac{10}{x-2}\right)=(x-2)\cdot\frac{5x}{x-2}$$

$$x(x-2)+10=5x$$

$$x^2-2x+10=5x$$

$$x^2-7x+10=0$$

$$(x-2)(x-5)=0$$

$$x-2=0 \quad\text{and}\quad x-5=0$$

$$x=2 \quad\text{and}\quad x=5$$

$x=2$ must be discarded since it must be excluded as an input.

30. Factoring permits us to simplify before multiplying.

Final Exam Review

1. **a.** $-7+(-5)=-12$

   b. $-3-(-8)=-3+8=5$

   c. $-6+11=5$

   d. $(-5)(12)=-60$

   e. $(-4)(-16)=64$

   f. $|4+(-8)|=|-4|=4$

   g. $\left|\dfrac{-45}{9}\right|=|-5|=5$

   h. $\dfrac{3}{2}\cdot\dfrac{4}{15}=\dfrac{\cancel{3}^{1}}{\cancel{2}_{1}}\cdot\dfrac{\cancel{4}^{2}}{\cancel{15}_{5}}=\dfrac{2}{5}$

   i. $\dfrac{-2}{3}\div\dfrac{5}{6}=\dfrac{-2}{3}\cdot\dfrac{6}{5}=\dfrac{-2}{\cancel{3}_{1}}\cdot\dfrac{\cancel{6}^{2}}{5}=\dfrac{-4}{5}$

3. **a.** $\dfrac{bcd}{bdf}=\dfrac{\cancel{b}c\cancel{d}}{\cancel{b}\cancel{d}f}=\dfrac{c}{f}$

   b. $\dfrac{-3rs}{12sx}=\dfrac{-\cancel{3}^{1}r\cancel{s}}{\cancel{12}_{4}\cancel{s}x}=\dfrac{-r}{4x}$

   c. $\dfrac{a^{3}}{a^{2}}=a^{3-2}=a^{1}=a$

   d. $\dfrac{x-2}{2-x}=\dfrac{x-2}{-1(x-2)}=\dfrac{\cancel{x-2}^{1}}{-1\cancel{(x-2)}_{1}}$

   $=\dfrac{1}{-1}=-1$

   e. $\dfrac{(4m^{2}n)^{3}}{6n^{2}}=\dfrac{4^{3}\left(m^{2}\right)^{3}n^{3}}{6n^{2}}=\dfrac{64m^{6}n^{3}}{6n^{2}}$

   $=\dfrac{\cancel{64}^{32}m^{6}n^{3}}{\cancel{6}_{3}n^{2}}=\dfrac{32}{3}m^{6}n^{3-2}=\dfrac{32}{3}m^{6}n^{1}$

   $=\dfrac{32m^{6}n}{3}$

f. $\dfrac{ab}{c}\div\dfrac{ac}{b}=\dfrac{ab}{c}\cdot\dfrac{b}{ac}=\dfrac{\cancel{a}b}{c}\cdot\dfrac{b}{\cancel{a}c}=\dfrac{b^{2}}{c^{2}}$

g. $n^{3}n^{4}=n^{3+4}=n^{7}$

h. $\left(mn^{2}\right)^{3}=m^{3}\left(n^{2}\right)^{3}=m^{3}n^{6}$

i. $m^{-1}m^{-2}=m^{-1+-2}=m^{-3}=\dfrac{1}{m^{3}}$

j. $\left(3x^{3}\right)^{0}=1$

k. $\dfrac{6x^{-3}}{2x^{2}y^{-2}}=\dfrac{6y^{2}}{2x^{2}x^{3}}=\dfrac{\cancel{6}^{3}y^{2}}{\cancel{2}_{1}x^{2}x^{3}}=\dfrac{3y^{2}}{x^{5}}$

l. $\left(3x^{-3}y^{2}\right)^{3}=3^{3}\left(x^{-3}\right)^{3}\left(y^{2}\right)^{3}$

   $=27x^{-9}y^{6}=\dfrac{27y^{6}}{x^{9}}$

5. **a.** $2x+3y=12$

   $3y=-2x+12$

   $y=-\tfrac{2}{3}x+4$

   b. $2x-3y=7$

   $2x=3y+7$

   $x=\tfrac{3}{2}y+\tfrac{7}{2}$

   c. $ax+by=c$

   $by=-ax+c$

   $y=-\tfrac{a}{b}x+\tfrac{c}{b}$

   d. $\dfrac{x}{x+1}=\dfrac{3}{8}$

   $8(x+1)\cdot\dfrac{x}{x+1}=8(x+1)\cdot\dfrac{3}{8}$

   $8x=3(x+1)$

   $8x=3x+3$

   $5x=3$

   $x=\tfrac{3}{5}$

e. $\dfrac{P_1 V_1}{T_1} = \dfrac{P_2 V_2}{T_2}$

$T_1 T_2 \cdot \dfrac{P_1 V_1}{T_1} = T_1 T_2 \cdot \dfrac{P_2 V_2}{T_2}$

$T_2 P_1 V_1 = T_1 P_2 V_2$

$\dfrac{T_2 P_1 V}{T_1 V_2} = \dfrac{T_1 P_2 V_2}{T_1 V_2}$

$\dfrac{T_2 P_1 V}{T_1 V_2} = P_2$

f. $3(x+2) = 7(x-10)$

$3x + 6 = 7x - 70$

$-4x + 6 = -70$

$-4x = -76$

$x = 19$

g. $4^2 + x^2 = 8^2$

$16 + x^2 = 64$

$x^2 = 48$

$x = \pm\sqrt{48} \approx \pm 6.9$

h. $d^2 = [4 - (-3)]^2 + (5-2)^2$

$d^2 = [4+3]^2 + 3^2$

$d^2 = 7^2 + 9$

$d^2 = 49 + 9$

$d^2 = 58$

$d = \pm\sqrt{58} \approx \pm 7.6$

i. $6 = \sqrt{3x-3}$

$6^2 = \left(\sqrt{3x-3}\right)^2$

$36 = 3x - 3$

$39 = 3x$

$13 = x$

7. a. $12xy + 3xy^2 + 6x^2 y^2$

$= 3xy(4 + y + 2xy)$

b. $x^2 + x + 1$ does not factor

c. $x^2 + x - 20 = (x-4)(x+5)$

d. $4x^2 - 9 = (2x+3)(2x-3)$

e. $6x^2 + x - 2 = (2x-1)(3x+2)$

f. $6x^2 - 11x - 2 = (x-2)(6x+1)$

g. $x^2 - 10x + 25$

$= (x-5)(x-5)$ or $(x-5)^2$

h. $6x^2 + 3x - 9 = 3(2x^2 + x - 3)$

$= 3(x-1)(2x+3)$

9.

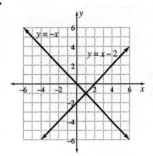

11. Substitute the second equation into the first equation to find y.

$3(6 - 2y) + 4y = 10$

$18 - 6y + 4y = 10$

$18 - 2y = 10$

$-2y = -8$

$y = 4$

Substitute $y = 4$ into the second equation to find x.

$x = 6 - 2(4)$

$x = 6 - 8$

$x = -2$

13. Multiply the first equation by 3 and the second equation by -4 and add the two equations to find x.

$9x + 12y = -60$

$\underline{8x - 12y = -8}$

$17x = -68$

$x = -4$

Multiply the first equation by -2 and the second equation by 3 to find y.

$-6x - 8y = 40$

$\underline{6x - 9y = -6}$

$-17y = 34$

$y = -2$

15. Substitute the second equation into the first equation to find y.

$$(8-y)+y=8$$
$$8-y+y=8$$
$$8=8$$

Infinite number of solutions

17. $m=\dfrac{5-(-3)}{2-2}=\dfrac{5+3}{0}=\dfrac{8}{0}=$ undefined

$d=\sqrt{(5-(-3))^2+(2-2)^2}=\sqrt{8^2+0^2}$

$=\sqrt{64}=8$

19. $m=\dfrac{-2-(-3)}{-5-2}=\dfrac{-2+3}{-7}=\dfrac{1}{-7}=-\dfrac{1}{7}$

$d=\sqrt{(-2-(-3))^2+(-5-2)^2}$

$=\sqrt{1^2+(-7)^2}=\sqrt{1+49}=\sqrt{50}$

$=\sqrt{25\cdot2}=5\sqrt{2}\approx7.1$

21. The input that must be excluded is $x=0$.

$$\frac{4}{x}=\frac{6}{x}$$
$$x\cdot\frac{4}{x}=x\cdot\frac{6}{x}$$
$$4=6$$

No solution

23. The input that must be excluded is $x=4$.

$$2x+\frac{3x}{x-4}=\frac{12}{x-4}$$
$$(x-4)\left(2x+\frac{3x}{x-4}\right)=(x-4)\frac{12}{x-4}$$
$$2x(x-4)+3x=12$$
$$2x^2-8x+3x=12$$
$$2x^2-5x-12=0$$
$$(2x+3)(x-4)=0$$

$2x+3=0$　　and　　$x-4=0$

$2x=-3$　　　　　　　$x=4$

$x=-\frac{3}{2}$

Therefore, only $x=-\frac{3}{2}$ is a solution.

25. The input that must be excluded is $x=-5$.

$$\frac{x}{1}=\frac{14}{x+5}$$
$$(x+5)\cdot\frac{x}{1}=(x+5)\cdot\frac{14}{x+5}$$
$$x(x+5)=14$$
$$x^2+5x-14=0$$
$$(x+7)(x-2)=0$$

$x+7=0$　　and　　$x-2=0$

$x=-7$　and　　　　$x=2$

27. There are no inputs that must be excluded.

$$\frac{2x}{5}-\frac{x}{10}=\frac{3x}{10}$$
$$10\left(\frac{2x}{5}-\frac{x}{10}\right)=10\cdot\frac{3x}{10}$$
$$4x-x=3x$$
$$3x=3x$$
$$0=0$$

All real numbers

29. The inputs that must be excluded are $x<-13$.

$$\sqrt{x+13}=5$$
$$\left(\sqrt{x+13}\right)^2=5^2$$
$$x+13=25$$
$$x=12$$

31. The inputs that must be excluded are $x<-13$.

$$\sqrt{x+13}=0$$
$$\left(\sqrt{x+13}\right)^2=(0)^2$$
$$x+13=0$$
$$x=-13$$

No solution since -13 is not less than -13.

33. a. $(4,0)$

b. $(0,2)$

c. $x=3$

d. $\sqrt{4-x}$ is undefined

e. The graph shows no negative values for y because $y=\sqrt{4-x}$ is always positive.

35. The person lost about 4 pounds in the first 12 days or so. Then the person went off the diet and started gaining weight.

37. a. $3^0 + 4^2 = 1 + 16 = 17$

    b. $5^2 + 5^{-1} = 25 + \frac{1}{5} = 25\frac{1}{5}$ or $\frac{126}{5}$

    c. $\sqrt{25} + 16^{\frac{1}{2}} = 5 + 4 = 9$

    d. $\sqrt{225} + 25^{\frac{1}{2}} = 15 + 5 = 20$

    e. $\sqrt{64x^2} = 8|x|$

    f. $\sqrt{64x^2} = 8x$

    g. indeterminate

    h. $\sqrt{x}$ = undefined for real numbers

39. $\dfrac{4x^2 - 1}{2x^2 + 5x + 2} = \dfrac{(2x-1)(2x+1)}{(2x+1)(x+2)}$

$\quad = \dfrac{(2x-1)\,\cancel{(2x+1)}}{\cancel{(2x+1)}\,(x+2)} = \dfrac{2x-1}{x+2}$

41. a. $\dfrac{\$6.00 \text{ per hour}}{3 \text{ rooms cleaned per hour}}$

$\quad = \dfrac{\frac{\$6}{\text{hour}}}{\frac{3 \text{ rooms cleaned}}{\text{hour}}}$

$\quad = \dfrac{\$6}{\text{hour}} \div \dfrac{3 \text{ rooms cleaned}}{\text{hour}}$

$\quad = \dfrac{\$6}{\text{hour}} \cdot \dfrac{\text{hour}}{3 \text{ rooms cleaned}}$

$\quad = \dfrac{\$\overset{2}{\cancel{6}}}{\cancel{\text{hour}}} \cdot \dfrac{\cancel{\text{hour}}}{\underset{1}{\cancel{3}} \text{ rooms cleaned}}$

$\quad = \dfrac{\$2}{\text{room cleaned}}$

b. $d = -\dfrac{1}{2}\left(\dfrac{32.2 \text{ ft}}{\sec^2}\right)(3 \sec)^2$

$\qquad + \left(\dfrac{22 \text{ ft}}{\sec}\right)(3 \sec) + 100 \text{ ft}$

$\quad = -\dfrac{1}{2}\left(\dfrac{32.2 \text{ ft}}{\sec^2}\right)\dfrac{(3 \sec)^2}{1}$

$\qquad + \left(\dfrac{22 \text{ ft}}{\sec}\right)\left(\dfrac{3 \sec}{1}\right) + 100 \text{ ft}$

$\quad = -\dfrac{1}{2}\left(\dfrac{32.2 \text{ ft}}{\sec^2}\right)\dfrac{9 \sec^2}{1}$

$\qquad + \left(\dfrac{22 \text{ ft}}{\sec}\right)\left(\dfrac{3 \sec}{1}\right) + 100 \text{ ft}$

$\quad = -\dfrac{1}{\underset{1}{\cancel{2}}}\left(\dfrac{\overset{16.1}{\cancel{32.2}} \text{ ft}}{\cancel{\sec^2}}\right)\dfrac{9 \cancel{\sec^2}}{1}$

$\qquad + \left(\dfrac{22 \text{ ft}}{\cancel{\sec}}\right)\left(\dfrac{3 \cancel{\sec}}{1}\right) + 100 \text{ ft}$

$\quad = -144.9 \text{ ft} + 66 \text{ ft} + 100 \text{ ft} = 21.1 \text{ ft}$

43. a. $\dfrac{2-x}{xy^2} \cdot \dfrac{x^2}{x-2} = \dfrac{\overset{-1}{\cancel{2-x}}}{\underset{1}{\cancel{x}}\, y^2} \cdot \dfrac{\overset{x}{\cancel{x^2}}}{\underset{1}{\cancel{x-2}}}$

$\quad = \dfrac{-x}{y^2} = -\dfrac{x}{y^2}$

b. $\dfrac{x^2 - 3x - 4}{x^2 - 2x} \cdot \dfrac{x^2 - 4}{4 - x}$

$\quad = \dfrac{(x-4)(x+1)}{x(x-2)} \cdot \dfrac{(x-2)(x+2)}{4-x}$

$\quad = \dfrac{\overset{1}{\cancel{(x-4)}}(x+1)}{x\,\cancel{(x-2)}} \cdot \dfrac{\cancel{(x-2)}(x+2)}{\underset{-1}{\cancel{4-x}}}$

$\quad = \dfrac{(x+1)(x+2)}{-x} = -\dfrac{(x+1)(x+2)}{x}$

c. $\dfrac{x-4}{x+2}+\dfrac{x^2-3x-4}{x^2-4}$

$=\dfrac{x-4}{x+2}+\dfrac{x^2-3x-4}{(x-2)(x+2)}$

$=\dfrac{x-4}{x+2}\cdot\dfrac{x-2}{x-2}+\dfrac{x^2-3x-4}{(x-2)(x+2)}$

$=\dfrac{x^2-6x+8}{(x+2)(x-2)}+\dfrac{x^2-3x-4}{(x+2)(x-2)}$

$=\dfrac{2x^2-9x+4}{(x+2)(x-2)}=\dfrac{(2x-1)(x-4)}{(x+2)(x-2)}$

d. $\dfrac{x^2+3x+2}{x^2+6x+9}\div\dfrac{x+2}{x+3}=\dfrac{x^2+3x+2}{x^2+6x+9}\cdot\dfrac{x+3}{x+2}$

$=\dfrac{(x+1)\,\cancel{(x+2)}}{(x+3)\,\cancel{(x+3)}}\cdot\dfrac{\cancel{x+3}}{\cancel{x+2}}=\dfrac{x+1}{x+3}$

e. $\dfrac{1}{3x}-\dfrac{2}{5x}=\dfrac{1}{3x}\cdot\dfrac{5}{5}-\dfrac{2}{5x}\cdot\dfrac{3}{3}$

$=\dfrac{5}{15x}-\dfrac{6}{15x}=\dfrac{-1}{15x}=-\dfrac{1}{15x}$

f. $8x\left(\dfrac{1}{2x}-\dfrac{3}{x}\right)=8x\left(\dfrac{1}{2x}\right)-8x\left(\dfrac{3}{x}\right)$

$=4-24=-20$

g. $3(x-2)\left(\dfrac{4}{x-2}+\dfrac{x}{3}\right)$

$=3(x-2)\left(\dfrac{4}{x-2}\right)+3(x-2)\left(\dfrac{x}{3}\right)$

$=12+x(x-2)$

$=12+x^2-2x=x^2-2x+12$

45. a. For $25, the payment due is $25.

For $35, the payment due is
$30+0.2($35-$30)$
$=$30+0.2($5)=$31$

For $95, the payment due is
$30+0.2($95-$30)$
$=$30+0.2($65)=$43$

For $100, the payment due is
$30+0.2($100-$30)$
$=$30+0.2($70)=$44$

For $105, the payment due is
$50+0.5($105-$100)$
$=$50+0.5($5)=$52.50$

b. For ($0,$30], $y=x$.
For ($30, $100], $y=30+0.2(x-30)$
For ($100, $+\infty$), $y=50+0.5(x-100)$

c. The brackets, [], include the endpoints and the parentheses, ( ), excludes the endpoints.

d. Yes, parentheses should be used with the infinity symbol, ∞.

e. $0<x\le30$

$30<x\le100$

$100<x<\infty$

47. a. 2.7564×10^9 miles

b. $\dfrac{4551.4\times10^6\text{ mi}}{1}\cdot\dfrac{1\text{ sec}}{186,000\text{ mi}}$

$=\dfrac{4551.4\times10^6\text{ }\cancel{\text{mi}}}{1}\cdot\dfrac{1\text{ sec}}{186,000\text{ }\cancel{\text{mi}}}$

$=24469.89247$ seconds

$\dfrac{24469.89247\text{ sec}}{1}\cdot\dfrac{1\text{ min}}{60\text{ sec}}\cdot\dfrac{1\text{ hr}}{60\text{ min}}$

$=\dfrac{24469.89247\text{ }\cancel{\text{sec}}}{1}\cdot\dfrac{1\text{ }\cancel{\text{min}}}{60\text{ }\cancel{\text{sec}}}\cdot\dfrac{1\text{ hr}}{60\text{ }\cancel{\text{min}}}$

≈6.8 hours

c. $\dfrac{200\text{ m}}{125.96\text{ sec}}\cdot\dfrac{60\text{ sec}}{1\text{ min}}\cdot\dfrac{60\text{ min}}{1\text{ hr}}\cdot\dfrac{1\text{ mi}}{1609\text{ m}}$

$=\dfrac{200\text{ }\cancel{\text{m}}}{125.96\text{ }\cancel{\text{sec}}}\cdot\dfrac{60\text{ }\cancel{\text{sec}}}{1\text{ }\cancel{\text{min}}}\cdot\dfrac{60\text{ }\cancel{\text{min}}}{1\text{ hr}}\cdot\dfrac{1\text{ mi}}{1609\text{ }\cancel{\text{m}}}$

≈3.55 mph

d. $\dfrac{28300\text{ }\cancel{\text{km}}}{\text{hr}}\cdot\dfrac{1000\text{ }\cancel{\text{m}}}{1\text{ }\cancel{\text{km}}}\cdot\dfrac{1\text{ mi}}{1609\text{ }\cancel{\text{m}}}$

$\approx17,600$ mph

49. a.

Length of edge of box, x	Volume of cube, x^3	Surface area, $6x^2$
10	1000	600
20	8000	2400
40	64000	9600
n	n^3	$6n^2$
$2n$	$(2n)^3 = 8n^3$	$6(2n)^2 = 6(4n^2) = 24n^2$

b. If we double the length of an edge of a box, the volume is 8 times as large.

c. If we double the length of an edge of the box, the surface area is 4 times as large.

d.
$$c^2 = 20^2 + 20^2$$
$$c^2 = 400 + 400$$
$$c^2 = 800$$
$$c = \sqrt{800} \approx 28.28 \text{ units}$$

50.

Family Size, x	Monthly Income, y
1	$507
2	$682
3	$857
4	$1032
5	$1207
6	$1382
7	$1557
8	$1732

For each unit increase in 'Family Size', the 'Monthly Income' increases by $175.

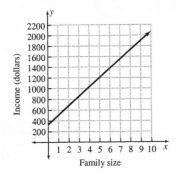

The graph would cross the y-axis at $332.

Rule:
The monthly income is equivalent to $332 more than the product of $175 and the number of people in the family.
That is, $y = \$332 + \$175x$.

27 in   3 ft

50 min   1.5 hrs

$$\frac{x}{14} = \frac{3}{5}$$ $$\frac{x}{3} = \frac{14}{5}$$